COMPLEX ADAPTATIONS
IN EVOLVING POPULATIONS

D0982275

T. H. FRAZZETTA

University of Illinois

COMPLEX ADAPTATIONS IN EVOLVING POPULATIONS

SINAUER ASSOCIATES, INC. ● PUBLISHERS
SUNDERLAND, MASSACHUSETTS

COMPLEX ADAPTATIONS IN EVOLVING POPULATIONS

First Edition

Copyright © 1975 by Sinauer Associates, Inc.

Printed in U.S.A.

Library of Congress Catalog Card Number: 74–24359

ISBN: 0–87893–194–5 (Cloth Edition); 0–87893–195–3 (Paper Edition)

To my parents

Contents

Preface

There was no noticeable beginning to the storm in the distance. Surely an hour ago the sky had been clear. Yet, a blackening blanket now stretched over the jungle as if it had always been there. The massive column of approaching rain seemed more a brace than a pendant to the dark expanse above it, widening as I watched it come, threatening to suffocate the entire Yucatán Peninsula.

Though my vantage point was above the tree canopy surrounding me for miles, no whisper of air, no wink of leaf anticipated the storm's arrival. I wanted very much to remain where I was through the storm, crouched against the top terrace of the Mayan temple. But the uneasiness of the hushed air was becoming contagious, making me feel vulnerable and unreasonably apprehensive. I looked up, wanting the retreating blue sky to give me a moment's reassurance. The sky was still cheerful above the stone-latticed roof comb which crowned the temple. I noted how hard it was not to be distracted by the magnificent ruin itself. The temple at Etzna seems not to attract the throngs of sightseers that are ever present at other Mayan edifices. I had it to myself. It rose starkly out of the midst of the jungle like a crumbling declaration. Abandoned perhaps more than half a millennium before the arrival of Columbus in the New World, it had mingled with the persistent jungle and, despite its startling suddenness among trees and vines, seemed now a natural part of the landscape.

A tiny scratching near my feet drew my gaze to meet a small, tan lizard, indifferently facing stormward as it perched on a flat stone. The storm. By now, after my mental meanderings, it was nearly overhead. An explosion of screams just below accompanied a flurry of green feathers. A flock of parrots in disorganized flight tore itself from the treetops, a part of the jungle itself wrenching free and flying away. With that, the first breeze worried the grass stems, then rose swiftly in strength, bringing cold dots of rain splashing on my face. Now great gray curtains were marching past, and through, hissing and rattling down on the forest canopy. I thought I could be sure that this was merely the overture to some indescribable fury yet to be unchained, but with anticlimactic prematureness the sun groped its way through the thick black clouds and ripped them to tatters. In witness of the great power of a storm or sudden burst of sunlight, the mind can struggle against its most available reaction, fear, to a state of almost raving gratitude for the mercy of continued existence. I felt foolish to have given myself to such extravagant anxieties for nothing more than a summer rain. A pair of rainbow arches intertwined across the dripping horizon. The sight seemed almost overdone; one rainbow would have been sufficient.

The surrounding elements were welded together by the glistening wetness they shared. The great forest, the magnificent structure of a bygone civilization with an exquisite sense of art and architecture, the lizard hiding somewhat unsuccessfully in a rock pile, myself—all these things stood for an instant together in time, frozen in all our tracks at some point in our evolution. The intricateness, the coordination, even the coincidence of all of it was suddenly immense, though in pace with the timeless rhythm of physics and chemistry which, for organic things, has molded the startling properties of life. The living things of the world are, to nearly all of us, the most intriguing and continuously amazing. Through them we see ourselves. Each of them has its remarkable capabilities—adaptations to ensure an interaction with its environment in a fashion that promotes its survival and the survival of its heirs.

It is enormously easy to be awed by the adaptations of living things, to be stunned by their complexity and suitability. But it should be easier than it seems to be to feel the evolutionary common denominator that all life shares with us; and to accept the inescapable but not humiliating fact that much of mankind can be seen in a tree or a lizard. And I do not doubt that some of it can be seen more easily there than in man himself. Despite the great diversity of living things, the sublime adaptations that to each of them are the equipment of life share similarities as striking as the differences. That the similarities can often best be appreciated through the differences is a paradox in syntax rather than in practice.

The evolution of major adaptations of animals is the subject of this book. Complex adaptations must have evolved according to principles compatible with the evolution of much simpler characters, but is this fact *sufficient* to the understanding of major adaptations? That is the question posed in these pages.

I have made no attempt to write an agreeable essay on the subject, and there are places where I have deliberately courted controversy on slender twigs of inference. But if subconsciously I have wanted a book of heresies, I shall have to learn to live with disappointment; in recent years a number of writers, each in his own way, have encouraged serious scrutiny of evolutionary considerations, especially where major, complex adaptive features are concerned.

I have tried to reach into several biological disciplines, in the hope of grasping enough connecting threads to weave a small fabric of diverse evolutionary phenomena which can be interrelated by a mutual, theoretical consistency. In doing this I have had no interest in producing a tiny treatise on morphology, on population biology, or on any other separately defined area of biology. The book has been designed for general reading on the topic of major, evolutionary transitions, as a text for courses in evolutionary topics, or as a companion text for several sorts of courses in distinct, biological disciplines.

It is my hope that the book will be readable for anyone interested in evolution who has had at least a general college-level biology course, with some elementary training in a few other scientific or science-related subjects. Although a quick flip through the book will reveal a few integral signs and so forth, I have tried to introduce all mathematical arguments through simple algebra. When I could, I avoided very specialized terminology, but there are places where this was difficult, and others where it was impossible.

I find myself tremendously indebted to a number of individuals who have provided the basics of encouragement, opinions, and resources. It is largely through the encouragement of my publisher, Andrew D. Sinauer, that this book was undertaken, and through his patience that it was completed. Edward O. Wilson added encouragement and provided continual, detailed suggestions. Leigh Van Valen made extensive comments and many constructive criticisms of the entire manuscript. I have benefited greatly from conversations with Dennis W. Nyberg, Bruce P. Hanna, Gilbert Waldbauer, Robert L. Metcalf, Mary F. Willson, Edward H. Brown, David L. Nanney, Alexa Clemans, Carlos Pinkham, Joshua Laerm, Clarice Prange, Kenneth Kardong, and Lewis Klein. The manuscript was typed by Jeanne Tinsley, Bonnie Bokszczanin, and Alexa Clemans. Many of the illustrations are the work of Carol Coope and Alice Prickett.

T. H. Frazzetta
Urbana, Illinois

1

ANIMALS
AS MACHINES

There was a time in my earlier days when the intrinsic beauty of animals would have been dulled by a comparison with machinery. Yet even a child can appreciate in a beast a sense of integrated wholeness, of design, of cooperation among its parts. These are, in fact, the very attributes that somehow give to animals their lovely and intriguing character. My reluctance to think of animals and machines in the same mental breath betrayed a fear that my treasured appreciation of animals would pale. I am sure that it removed animals from the too-familiar world of noisy, boastful, and often meaningless contrivances of human ingenuity.

Today I can no longer deny the connection between animals and machinery. But almost unexpectedly I discovered that I have lost none of my admiration and wonder of animals, nor any of my respect for them. On the contrary, they hold their private mysteries of evolution and diversity of form and resist any tarnish from the exquisitely mundane, man-made devices to which they can be compared so handily.

An animal is a highly integrated machine and, because it is, it is convenient rather than analytical to regard it in pieces, as a collection of separate characters and adaptations. Such an approach, if carried very far, will result in too narrow a focus on individual characters to recognize their close interrelationships with one another.

Even so, simplification achieved through that focus is not without splendid utility. Much of our conceptual understanding of evolution has been provided by the study of individual characters by population geneticists and systematists.

The concepts derived from the study of characteristics treated individually are customarily extended to explain all evolutionary phenomena. Too often the extension is made with a vague confidence that what is true for simple, isolated characters must be equally and also sufficiently true for sets of interrelated features. I am suspicious, however, that an understanding of evolution taken from consideration of individual characters might be a bit different from an understanding of the evolution of complex, integrated systems. If I properly appreciate the views of a number of modern evolutionists, I need not feel terribly alone in my suspicions. In fact, much of what I wish to say in this book has been said before. It is my hope in these pages to distill the essence of some of these viewpoints.

Every animal, no matter how simply constructed, has a nearly incredible organization among its parts. This is true at every structural level, from the biochemical to the macroscopic. Each of the functioning systems that together make up the animal contains elements at many structural levels. Often it is possible to imagine parts of several functionally interdependent systems to be sorted out in parcels that relate closely to a single aspect of the animal's requirements for life and reproduction. Although somewhat inaccurate, it is very convenient to deal with each parcel as if it were discrete and to think in terms of separate adaptations for feeding, for locomotion, for respiration, and so forth. Each adaptation can be envisioned as being subdivided into even finer adaptations. But unlimited subdivision will eventually get us to consideration of single characters in isolation, so the subdividing must be halted at some point if we are to be left with an adaptation that is complex.

A complex adaptation is one constructed of *several* components that must blend together operationally to make the adaptation "work." It is analogous to a machine whose performance depends upon careful cooperation among its parts. In the case of the machine, no single part can greatly be altered without changing the performance of the entire machine. There is little difficulty in providing a specific illustration of the machine analogy. Figure 1 shows one of several kinds of straight-line mechanisms. A frequent requirement in many machine linkages is the motion of certain parts in a straight line rather

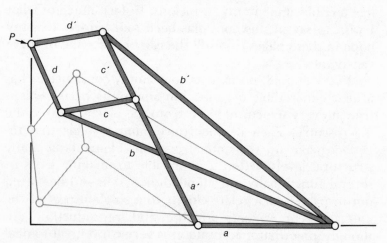

1 Peaucellier straight-line mechanism. The grey image shows the mechanism in a displaced position. The point *P* always traces a straight line along the ordinate axis.

than in the curve of an arc. The Peaucellier straight-line mechanism, the example chosen, is something of an engineering dinosaur in that it is scarcely ever used nowadays; other, more easily constructed straight-line devices enjoy far greater favor with manufacturers [1].* Despite its practical limitations, the Peaucellier machine has advantages for our purposes.

The mechanism consists of eight links joined together in a manner to allow rotation of each link about one or more other links. One of the links (a) is attached to and lies along the abscissa of two mutually perpendicular axes. I have designated as P the midpoint of the joint connecting links d and d'. The ordinate axis has been constructed through point P for a very special reason. Any motion of the mechanism which involves the swinging of link a' to the left or right produces straight-line motion of P along the ordinate.

Of course, if P is to move in a straight line, there are certain rules of construction that the mechanism must follow. It is necessary that all "pairs" of links be of equal length. Thus if L_a and $L_{a'}$ are the lengths of links a and a', the expression $L_a = L_{a'}$ represents one of the conditions required for straight-line motion of P. This is not, of course, the only requirement. The total array of requirements is: $L_a = L_{a'}$, $L_b = L_{b'}$, $L_c = L_{c'} = L_d = L_{d'}$. But these requirements only designate the conditions necessary to enable P to move in a straight line perpendicular to the abscissa; they do not guarantee that the path of P will follow any *particular* line perpendicular to the abscissa. For example, suppose that we imagined each link in

---

*Notes, indicated by numbers in brackets, are gathered at the end of each chapter. References are at the back of the book.

Figure 1 to become shorter but that we retained the equality of lengths between pairs. Point *P* would still trace a straight line (since paired links are equal), but the line would lie to the right of the ordinate shown.

Now let us use this example plus one assumption to illustrate something of complex adaptations. For our assumption suppose that the task required of the straight-line device is such that *P must* move along the ordinate depicted in Figure 1. Immediately it is obvious that we cannot alter the length of any link without impairing the mechanism's performance. The components of the system, the links, are functionally integrated, and a change in the form (length) of any of them will upset their mutual, operational harmony.

But suppose that an engineer sees a way to improve the entire machine of which our straight-line device is a part. And suppose that this "evolutionary change" must not disturb the path of *P* but does require modification of the length of link *a*. The problem is readily solved: any change in *a* must be matched by an equal change in *a'*, and compensating changes must be made in one or more of the remaining pairs of links that will allow *P* to remain on its present ordinate axis. Although the solution to the problem is readily perceived by the engineer, the changes required are somewhat complicated and quite precise: no less than four links must be altered if any one is changed, and the alteration of any pair with respect to that of another pair is uniquely specified. In fact, if we maintain that the present path of *P* must be retained if a machine of the type considered is to perform properly, no change in the straight-line device is possible without correlated changes, occurring simultaneously, in a number of the constituent components.

Examples of complex adaptations in animals are a bit

harder to discuss than engineering devices. It is not because a large number of animal examples do not come readily to mind: the wing of a dragonfly, with its mechanically sensible venation and shape; the vertebrate eye, whose parts join in a delicate orchestration of function; and the swimming mechanism of a fish (which includes nearly the entire fish) are examples eager for consideration as complex adaptations. Still, artificial machinery is more easily analyzed. For the very reason that it is man-made, its sense of purpose and design is immediately available to a human mind. While often it is not difficult to appreciate the effect of slightly altering one or more components in a man-made machine, this appreciation is frequently dimmer when applied to animal adaptations.

Not only are artificial machines better understood from the standpoint of purpose, they are also simpler than biological devices, which may include a plethora of very different materials and have vital functions that range from molecular to visible levels.

The biological example I have chosen for illustration is a portion of the jaw apparatus of a python. The portion to be considered is fairly complex in that several components are involved. Yet it belongs to a larger apparatus from which its separation for illustrative convenience is, in truth, fairly unrealistic [2].

Like many other snakes, a python ambushes such prey as mammals and birds by a very rapid lunge of the head and forebody to secure its meals by means of its many long teeth. The teeth are recurved so as to hold relatively large, struggling prey in the jaws. The recurvature of the teeth, although well suited for holding prey once caught, slant the "wrong" way to ensure penetration of the prey during the lunge. Penetration would be more certain if the pointed tips were directed toward the prey, in the same direction as the movement of the head during the lunge. Actually, pythons employ two methods that direct the tooth tips to the front when prey is seized. The head is thrown far back on the neck; at the same time, the skull bones to which the teeth are attached are tilted to further point the teeth forward. Figure 2 shows

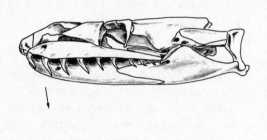

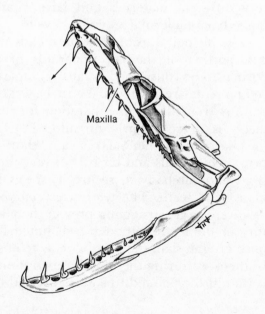

Maxilla

how the tooth tips are changed in direction as the lunge progresses. Skulls of snakes, and incidentally of many other vertebrate animals, are unlike our own mammalian skulls in that many upper jaw bones are very movable, like the links of a complicated machine.

After prey is seized it is rather quickly dispatched by

Attack lunge of a python. Just prior to the attack the mouth is **2** closed and the head oriented normally on the "neck." During the strike, the head is thrown backward as the mouth opens, while the maxilla is rotated upward. These movements direct the tooth tips forward toward the prey. The arrows show the tooth-tip directions. *Note on Teeth:* In this and the following two figures the teeth shown are those which are firmly anchored to the jaw bones. Snakes possess continuous, alternate tooth replacement throughout life. While some teeth are becoming loosened, in preparation for being shed, alternate tooth sites are receiving new teeth, which are becoming firmly attached to the jaw bones.

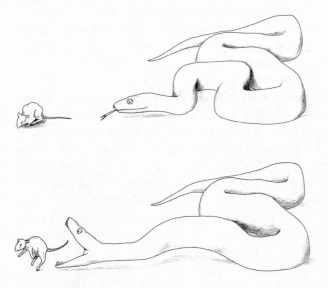

constriction, the python enveloping the prey in a coil tight enough to interfere with normal heart and breathing movements. Swallowing follows and is accomplished with the aid of the movable jaw bones. Their loose connections permit distension of the gullet to accommodate the impressively large prey that snakes often eat. During swallowing, the teeth of one side of the head hold the prey while those of the other side are withdrawn to allow the disengaged jaws to reach over the prey and secure a new grip. The process continues with each side of the head alternating engagement and disengagement while the jaws reach for successively new grips farther forward on the prey until, at last, the prey is drawn completely into the esophagus.

Several upper jaw bones bear teeth and function in both capture and swallowing of prey. Perhaps the most prominent are the pair of *maxillae* (shown in Figure 3), with their seventeen or so teeth apiece. Movement of each maxilla augments the directing of its tooth tips forward during the lunge and permits alternate engagement and disengagement of the teeth during swallowing.

Maxillary motion depends very largely on several other bony structures in the python's muzzle region (Figure 3), and all these will form the components in this example of a complex biological adaptation. The very toothy maxilla is suspended from the braincase part of the skull by the prefrontal, a very important bone that coordinates several kinds of movement. Prefrontal and maxilla are loosely connected by flexible tissue, and the maxilla can tilt and swing relative to the prefrontal. The prefrontal's very intricate shape relates to its function as a coordinator of movement. To simplify matters, we shall ignore not only much of its morphologic intricacy but also several of the movements that it coordinates.

The prefrontals are movably attached to the frontal

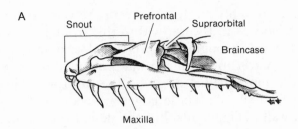

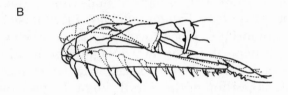

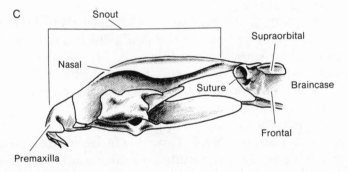

A. Muzzle bones in the African python. B. The dotted image **3** provides a simplified illustration of bone movements. C. Snout and its attachment to the braincase. Other muzzle bones have been removed.

bones. The pair of frontals encloses a portion of the brain, and their side walls are deeply recessed toward the midline of the skull to accommodate the eyes. It is against this recessed surface of each frontal that each prefrontal bone moves. The surface slants inward, from the top of the frontal, at roughly 45 degrees from the vertical. The prefrontal pivots as if it were held to this frontal surface by a pin driven through the prefrontal and into the frontal wall. As the prefrontal rotates about this imaginary pin, the most medial edges of the prefrontal slide against the frontal wall. These edges are actually tied to the frontal wall by ligaments, but the tie is loose enough to allow sliding motion of the bones, yet secure enough to give strength and stability to the prefrontal–frontal joint.

The upper part of the frontal wall joins a thin flake of a bone, the supraorbital, which forms a roof over the eye. Anteriorly this bone dips downward slightly, separating from the frontal wall to leave a small space. The space accommodates a portion of the dorsal rim of the prefrontal, and hence the supraorbital provides flexible support for the top of the prefrontal.

Movement of the prefrontal is brought about by muscles in the posterior half of the skull (which is outside the region of our present consideration). The effect of those muscles is to shove the maxilla forward. This, of course, pushes the prefrontal, which then rotates, carrying the maxilla forward and upward.

The maxilla tilts upward, pivoting around its attachment with the prefrontal. How much the maxilla pivots depends upon the magnitude and direction of the push against it from behind and the amount that the prefrontal rotates. There are elastic sheets of tissue running between prefrontal and maxilla which tend to pull the front of the maxilla up, together with the rotating prefrontal.

Lifting and tilting of the row of maxillary teeth is accompanied by movements of the snout. The snout rotates upward to provide freer operation of the prefrontal with less stretching of the overlying skin, and to allow greater ease in maxillary tooth penetration. Several bones make up the snout. Beyond rounding out the contours of the head, the snout houses sensory structures, is traversed by paired passages that conduct air from the nostrils, and exerts a little control over some maxillary movements. Moreover, in pythons, the most anterior snout bone (the premaxilla) bears several moderate-sized teeth.

The snout lifts by rotation about a snout–frontal joint. The joint involves the upper front faces of the two frontals, at the midline where they come together. There is, in fact, a very clear division in each frontal, marked by a suture connecting two edges of the bone. The suture divides each frontal into an upper and a lower portion, and it is with the upper portion that the snout is movably joined. Flexible connecting tissues tie frontals and snout together in a manner to permit snout rotation.

The primary agents lifting the snout are its ligamentous connections to the prefrontals. When the prefrontals rotate, the ligaments tug on the snout and pull it upward. Hence, as the prefrontals move, both snout and maxillae swing upward.

In most species of pythons the snout lifts about as much as does the maxilla, which is functionally reasonable if lifting of the snout is to provide clearance for penetration of the maxillary teeth. And in most species the maxilla lifts about as high as the anterior maxillary teeth are long, which is also reasonable if one considers that maxillary movements are the primary means by which the teeth are embedded and withdrawn during capture and swallowing of prey.

The interlocking function of several components qualifies the python muzzle as a complex adaptation. It is clear that we cannot significantly alter any one component without changing the way that the whole apparatus works. But how *much* of an alteration of any one component would be significant? For reasons already discussed, we cannot answer this question with exactly the same satisfaction that is possible when we are dealing with the Peaucellier mechanism. Although it is easy to admire what seems to be the perfection of organic devices, it is a lot harder to understand them thoroughly.

One way in which we may attempt to appreciate the interrelationships of components in a complex adaptation is to observe their evolutionary changes. Of course, this approach is not innocent of weaknesses, not the least of which is difficulty in documenting the course of events in the change itself. And there is always the problem that the known effects of evolution reveal merely what has happened—not what could or could not have happened. Caution, as always, is therefore necessary. However, reasonable caution should never become an excuse for an immobilizing fear, and the method of interpreting evolutionary change in complex adaptations can provide functional suggestions and hints not accessible otherwise.

The sort of muzzle just described is rather typical of most of the score or so python species alive today. This and other evidence suggests that the ancestors of present-day pythons possessed a very similar muzzle apparatus. Nevertheless, three closely related living pythons have departed significantly from this pattern. The most bizarre of these is *Chondropython viridis,* the emerald python, a lovely, bright-green inhabitant of the treetops of New Guinea and northern Australia. Its nearly totally arboreal existence contrasts with the life style of most other pythons and, as expected, is reflected in many aspects of its structure. In general build it is less heavy-bodied than most of its terrestrial relatives, a presumed advantage for moving about on slender branches. The jaw apparatus is remarkably flexible, for there are several features of an arboreal life that place a high adaptive priority on exceptionally movable jaw bones.

Unlike terrestrial snakes, *Chondropython* must search and stalk its prey in the very restrictive circumstances of the branches of a tree. A lunge at prey from a branch is likely to be less accurate than a lunge made from the broader, surer surface of the ground. The prey itself, if available in the treetops, whether it be a volant bird or an arboreal mammal, is undoubtedly fairly expert at maneuvering in three dimensions. And once the prey is secured by the jaws, it must be held, struggling, as the snake's head draws it back through the air to its waiting coils. Constriction in a tree must be more difficult than on the ground, and even after the prey is dispatched, it cannot be accidentally released from the jaws during swallowing, for if it is dropped, it is lost.

A jaw apparatus with exceptional abilities for nimble motion and for snaring prey seems especially prescribed. At the same time, the lighter build of the snake's entire body results in greatly lowered collision forces between the jaws and the prey during capture. A more flexible jaw construction (which is weaker) is thus permissible, and so are longer, relatively more slender teeth.

The muzzle region of *Chondropython*, shown in Figure 4, has a decidedly more formidable appearance than the muzzle of terrestrial pythons. Noticeable are the much longer, less recurved teeth. The reduced curvature seems to be a means to bring the tooth tips—especially those of the longer, anterior teeth—farther forward, toward the front of the jaw apparatus. With the tooth tips included among the more anterior elements of the jaws, their penetration into the prey during the lunge is made more

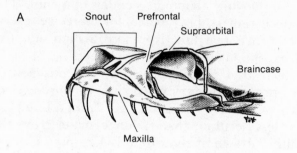

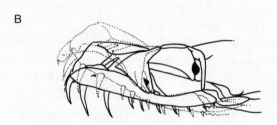

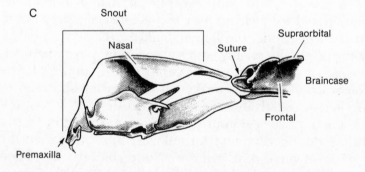

4 A. Muzzle bones in the green tree python *(Chondropython)*. B. The dotted image provides a simplified illustration of bone movements. C. Snout and its attachment to the braincase. Other muzzle bones have been removed.

probable. Of course, the reduced tooth curvature is less effective in holding prey once caught, but increased tooth length probably compensates by allowing deeper penetration.

Longer teeth could be an adaptive liability without a sufficiently movable maxillary bone. Movement of the maxilla depends largely upon the prefrontal's movement, and so we must examine the prefrontal with particular care.

Prefrontal movement in *Chondropython* is freer and greater in amount, thanks to the modified form of the prefrontal–frontal joints. The side wall of each frontal, which in terrestrial pythons slopes downward toward the skull's midline at about 45 degrees, is more nearly horizontal in *Chondropython*. It is as if this part of the frontal had been pushed inward and upward to form more of a ceiling and less of a wall in the orbital region. This has, of course, changed the plane of the prefrontal–frontal joint. The more nearly horizontal plane of the joint means that when the prefrontal lifts, its motion at the joint will involve less sliding, but more hinge-like swinging. The hinged joint with its freer movement is a mixed blessing—it is weaker, less able to resist dislocation than the prefrontal–frontal joints of other pythons. In compensation for this weakness the area of contact between the bones is increased, and this has been done by extending the joint region out laterally, over the orbit. But lateral extension of the frontal must occur by "invasion" of the territory primitively occupied by the supraorbital bone. Indeed, in *Chondropython* the supraorbital is smaller than in many other pythons, the broadened frontal having forced its reduction.

A corollary to these modifications is the increased size of the orbit. Arboreal snakes tend to be very visual animals, with eyes proportionately large. The changes favoring increased prefrontal mobility seem to be the same ones that permit a larger eye.

Horizontal flattening of the frontal side walls has an effect on the snout–frontal joint. As the frontal walls become pushed inward and upward, the lowermost and median parts of the walls are pressed upward accordingly. These lowermost and median parts include the frontal portions that lie *below* the suture that is present in the region of the snout joint. In most pythons the *upper* frontal portions—those lying above the suture—meet the snout to form a movable joint. But as the frontal walls, and thus the lower frontal portions, were pushed upward during the evolution of *Chondropython*, the upper frontal portions seem to have become horizontally thinner. Eventually the snout joint was shifted to the lower frontal portions, away from the now insubstantial upper parts.

Lengthening of the maxillary teeth ushers in other difficulties and requirements. As the teeth lengthen, closing the mouth becomes a problem, because the tips of the maxillary teeth tend to be exposed below the edge of the lower jaw. *Chondropython* deals with this in simple fashion—the maxilla is arched upwards in the muzzle region so that the tips of the long teeth are carried higher relative to the lower jaw.

The upward arch of the maxilla is accommodated by the high profile of the snout. In effect, the snout has been bent downward in the front and warped upward in the middle. I am not at all certain about the functional implications of this snout shape, but I can note that the bent snout is less prominent, protruding less in front. In this way, the long maxillary teeth become relatively more anterior and are more likely to be the first part of the snake's anatomy to make contact with the prey.

These changes, presumed to have occurred in the evolution of one adaptive form from another, hint generously of a fairly tight integration among components. There is a definite suggestion that certain modifications cannot greatly precede or lag behind others but must keep pace if the performance of the machine is not to become sloppy. Of course, much of our view of the transition and its underlying selective factors is inferential. A change in our interpretations could "loosen" the apparent tightness of interlocking functions. But at least some of our interpretation seems fairly sound; besides, we have had to deliberately ignore many components that probably integrate with those relatively few considered in the example.

We can only guess how long it took to make the evolutionary transition to *Chondropython* from a less-specialized ancestor. The fossil record, although not silent on the matter, is far from explicit. Snakes seem to have emerged in the late Cretaceous, and the larger python-like species only begin to show themselves in the Eocene. There are then, perhaps, as many as sixty or so million years available, although the actual change might have occurred in a fraction of that time.

Cases of evolutionary transitions that are better documented in the fossil record than that of *Chondropython* often involve many millions of years, whereas in the realm of man-made devices, major inventions and innovations occur in terms of centuries, decades, and, during rare flashes of intuition, instants. There are some very fundamental differences, however, between constructing a complex apparatus and evolving one. When an engineer designs new machinery, he consciously (and no doubt subconsciously) seeks to accumulate and concentrate relevant data and ideas from varied sources. Much of this information represents significant portions of the lifework of other engineers. There is thus available to him a tremendous array of potentially workable ideas and pertinent knowledge.

In the case of evolution by natural selection, the array of choices is usually considered to be less goal-directed, less workable, and less likely to include viable solutions to any adaptive problem, let alone the one at hand. The term "random variation" is often used to describe this array, to emphasize that the appropriateness of any one variant to the evolution of a particular, new adaptive requirement is purely a matter of luck.

Another difference is that when modifying the design of a machine, an engineer is not bound by the need to maintain a real continuity between the first machine and the modification. The new machine may be as different from the old as the muzzle of *Chondropython* is from that of most other pythons. But in evolution, transitions from one type to the next presumably involve a greater continuity by means of a vast number of intermediate types. Not only must the end product—the final machine—be feasible, but so must be all the intermediates. The evolutionary problem is, in a real sense, the gradual improvement of a machine while it is running!

Possibly the greatest advantage enjoyed by the engineer is his ability to learn from mistakes, his and those of others. A machine that fails, a ship that sinks, a plane that refuses to fly are all steps toward eventual success of a new engineering design. Each engineering failure can represent an unplanned experiment. As much has probably been learned through human failure as through success. And engineers can work out the "bugs" in devices that are nearly, but not quite successful.

Evolutionary failures are quite another matter. Needless to say, they teach nothing. An inferior organic modification will tend to be selected out of a population and leave nothing behind on which future generations can build. This is true even when the new modification is structurally but a breath away from an adaptive form that would be selectively favored.

The apparent disadvantages on the side of evolution have not prevented the appearance of millions of kinds of complicated animals. A recent book has the ominous title, *Mathematical Challenges to the Neo-Darwinian Interpretation of Evolution* [3]. In it Murray Eden presents thoughts on the complexity of living things and the enormous problems of evolving that much complexity. His critics (whose sentiments appear in the same book) argue with equal conviction for the adequacy of the unembellished, current evolutionary theory (which is, by the way, at least forty years old) to explain calmly the most intricate of adaptations. Undoubtedly I am revealing some hideous personalia to confess that I am attracted by both views in this controversy. The resulting dissonance leads me to expect, perhaps to hope, that both views are in some way basically compatible.

One of the major considerations in the evolution of organic complexity is time. Eden illustrates this in some highly simplified numerical exercises that dramatize these evolutionary problems at the molecular level. It is not difficult to devise yet another numerical game that will apply to the macroscopic level in terms of the kinds of adaptations that we have been examining.

The "game," since it delivers us to a conclusion at variance with the facts, is obviously played according to faulty rules. The point is, however, to identify the flaws and to find more appropriate rules to take their place. The most obvious flaw is the overwhelming simplicity of the assumptions, leading immediately to omissions of many details that could well be significant. In an attempt to counter that problem, the assumptions have been deliberately weighted against the conclusion that finally emerges. Like all such exercises, this game is instantly vulnerable if it is mistaken for anything but a numerical parable.

Among the many, major evolutionary events documented in the vertebrate fossil record is the rise, from primitive fishes, of the first land-living tetrapods. The earliest amphibians made their first appearance in the Upper Devonian. Their finned ancestors, the crossopterygian fishes, entered the fossil record in the Lower Devonian. Since the duration of the entire Devonian is about sixty-five million years, the transition from fish to amphibian could not have taken longer. It probably took fewer years than the total available.

Despite the wealth of morphological likenesses between the amphibians and crossopterygians, the number of modifications needed by the terrestrial pioneers must not have been small. Changes in the locomotion, respiratory, body support, integumentary, alimentary, circulatory, and behavioral systems must have been necessary. Within any single system there were probably required modifications of many parts. It seems reasonable that many of these modifications involved highly integrated components, such that certain changes would be inadaptive without a concurrence of certain other changes. In consideration of these factors it might be rashly conservative to state, as a first assumption, that a concurrence of only half a dozen particular modifications was sufficient to be favored by selection. We shall accept further the extreme assumption that with these six modifications present, selection was so favorable that the entire population could acquire this combination of traits within one or two generations. After this acquisition, an additional six modifications appeared in the population, which again had to occur in the same individual to be selectively advantageous, and so on, until at last we have an amphibian rather than a fish.

To make things simpler still, we shall not wait for mutational events to produce the required six characters in each step. Instead, assume that at every step the six features in question are each produced by but a single, dominant gene already present in the population. The frequency of each such gene is 10^{-3}, "common" by the standards of many population geneticists for new genes arising in populations. We shall ignore any genetic linkage effects and assume completely independent assortment. Also we shall assume the population to be constant over the span of time considered, with a rather high "selection-age" number [4] of 10^7. Certainly we are being generous when we assume that year-old individuals were reproductive, and we shall assume a generation time of one year.

Each diploid individual has two chances to possess each of the six genes, and his chances of combining by mating the genes governing all six features could be no better than $(2 \times 10^{-3})^6$. Since 2 is very nearly $10^{0.3}$, the chance of possessing all six genes is $(10^{0.3} \times 10^{-3})^6$, or $10^{-16.2}$. With a population containing 10^7 selection-age individuals, the combination will occur with a frequency of $10^{-9.2}$, or about once in every one billion years [5]. By the methods of our game, then, the attainment of merely the first step in the evolutionary transition—that involving only an initial set of six characters—will probably occur only once in a period of time far exceeding the length of the Devonian.

Something is very wrong with this because amphibians

did, after all, make an evolutionary appearance, as did many other animal types. This evolutionary game playing can be of benefit in supplying our dissatisfactions, uncertainties, and anxieties regarding current theory with the beguiling firmness of actual numbers. If the game results do not exactly define the problem, they lie at least *near* the problem that seems to haunt the confidence of many evolutionary biologists.

We need to dissect the assumptive framework of the game, provide corrections and additions to it, and seek alternative approaches consistent with discoveries in diverse biological areas. This cannot be attempted just yet; there are many things that must first be considered[6].

NOTES

1. Perhaps the simplest straight-line mechanism is the slider crank, which utilizes a block or "slider" that moves on a straight surface and is movably connected to a simple linkage. There are also stranger, and hence more fascinating devices, which, however, are not much more useful than the Peaucellier apparatus. The Watt straight-line linkage is one such—beautiful in its simplicity but imperfect in result; the line its reference point traces is not precisely straight. This does not much matter because for the purpose it was put to, the deviation from a truly straight path was not critical.

2. A more complete view of python jaws may be found in Frazzetta (1966).

3. Moorhead and Kaplan (1967).

4. I shall use "selection-age" numbers, as it seems more meaningful for our purposes than other measures of population size, such as "breeding size." Breeding size estimates how many members of a population have survived to the age of

reproduction and are thus poised, ready to make contributions of offspring to future generations. Characters affecting survival may become significant at any stage in an individual's life (including the very earliest stages); the portion of an animal population we shall be concerned with will often depend upon the age at which an individual will feel significant selection pressure. It is possible to imagine a statistically adjusted age at which a certain character (or character complex) will first have significance for natural selection. The number of individuals so aged in the population is the selection-age population size. After this age population members with relatively inferior characteristics will be reduced in proportion to others. Unless the characters in question do not reveal themselves to selection until sexual maturity, individuals attaining breeding age will contain a higher proportion of members bearing favorable traits; in terms of this simplified approach, by the time an average individual is expected to reach breeding age, his contest with selection pressures affecting many of his survival characteristics will be over.

5. A more accurate way to accommodate a diploid animal in terms of this evolutionary time game is through use of a simple Hardy–Weinberg distribution, although the method of approximation given in the text is quite adequate unless the gene frequencies are very high. If we are considering n_i characters that must occur simultaneously, an equilibrium distribution of genotypes will be given by

$$(A + a)^2 (B + b)^2 (C + c)^2 \cdots (I + i)^2. \qquad (1)$$

The capital letters represent the frequency of genes producing the special characters we are interested in; the lowercase letters are the allelic alternatives. In our example, $A = B = C = \cdots I = 10^{-3}$, and $a = b = c = \cdots i = 1 - 10^{-3}$. The equilibrium array of genotypes will occur and be perpetuated in the population after the initial period of chromosome disequilibrium (if any) has passed.

The squaring of values in any set of parentheses in formula (1) will give identical numerical values with respect to each set of allelic alternatives; and each squared term in parentheses will thus be equal to $A^2 + 2Aa + a^2 = 1$. Since we are only interested in the genotypes carrying characters designated by capital letters, for each set of squared parentheses we care only about those terms whose numerical value is equal to $A^2 + 2Aa < 1$.

Thus the array of genotypes, bearing at least one each of the genes for the special six characters, has a combined frequency

$$P = (A^2 + 2Aa)^{n_i}.$$

In logarithmic form,

$$\log P = n_i \log(A^2 + 2Aa).$$

If we dealt with just slightly more than six characters, the chance of combining the full complement would be hopelessly small. For eight genes at frequencies of 10^{-3}, the combination would be expected but once in more than 100 trillion years. If the eight genes were much more common, 10^{-2}, the time involved would still be more than one million years. But if a dozen genes with frequencies of 10^{-2} were considered, the combination of all twelve would be expected once in ten trillion years. Very common genes with frequencies of 10^{-1} would provide figures of only forty or so years with a dozen characters. But two dozen genes at the same frequency of 10^{-1} would combine once in ten billion years.

Obviously the structure of the game makes complex evolution highly improbable unless very few characters are adaptively integrated and dependent upon one another, or unless the genes of the new characters are exceedingly common.

6. The "game" will be considered directly in Chapter 4, and succeeding chapters will continue to touch on its implications. Those readers who are curious to know whether their

and my interpretations of the game are similar (plus unkind readers who wish to frustrate my organizational plans) will probably turn now to Chapter 4.

2

EVOLUTION
BY SMALL STEPS

With each passing year the once rather simplistic views on evolution continue to crumble, yet not a very long time ago general writings on evolution suffered from a sort of schizophrenia. The portions of many evolutionary discourses dealing with selection in populations pursued the subject in one manner, while the portions concerned with major complex adaptations approached their subject in a way that received little benefit from population considerations. In those days population geneticists were primarily interested in the short-term effects of selection, and cared only sporadically about the mechanisms of evolving complex systems. Morphologists and paleontologists did not fully recognize this, apparently. They could regard the most intricate of complex adaptations and, with a confident gesture toward population genetics, assure themselves and their audience that the findings of the population biologists somehow swept away all problems and concerns about how complex systems evolve.

Although things are vastly improved and improving, there are still traces of schizophrenia. There still remains some "format thinking" when approaching evolutionary subjects, especially if they touch on matters outside the immediate biological province of the thinker. Such problems are now being corrected. Not only are important questions being faced by a number of scientists, but there is growing admission that evolutionary aspects of complex adaptations have indeed been left dangling and that they need now to be rooted in the pertinent biological areas. At times the admission is a little self-conscious, especially when evolutionary biologists whisper grimly among themselves that they are disturbed by some of the pamphlets that attack evolution which are produced by fundamentalist religious groups. A certain number of these antievolutionary writings seem fairly well known to my colleagues.

The attack of the more sophisticated critics of evolution is invariably aimed at the narrow isthmus that connects the substantial bodies of evolutionary evidence—population considerations on one side, paleontological fact on the other. Although the alternative views proposed by these critics are so lacking in realism that they disqualify themselves, it clearly behooves those evolutionary biologists who feel uneasy about the attacks to widen the isthmus or to overhaul the theory.

The present chapter deals with several aspects of primarily traditional approaches to evolutionary mechanics. At this level of approach is found the observations and experiments on populations changing under selection's guidance. The traditional approach forms a reasonable point of departure for almost any sort of evolutionary discussion. Most ideas regarding the workings of natural selection have come through experiments, observations, and disciplined thinking about single characters or simple character systems. After all, that is the easiest way to come by such ideas. But the convenience may exact a price. More complicated systems could have evolutionary implications peculiar to themselves. If they do, evolutionary concepts based on simple character systems alone might lack proper balance.

Numerous published works call attention to the need to give special regard to complex character systems. The writings of De Beer, Eden, Goldschmidt, Gould, Mayr, Rensch, Sondhi, and Waddington, among a great many [1], have identified the need to regard complex systems in terms of integrated character groups. Although complex characters require special kinds of attention, there is no obvious reason why their evolutionary mechanics should not be *compatible* with those of simpler systems. Many, perhaps most, writers on this subject do not suggest otherwise. Yet there are some authors (e.g., Goldschmidt) who deny, outright, any such compatibility and others who seem only to hint at such a denial, but their inexplicitness leaves me uncertain of their real feelings.

Although it is true that a great many papers on varied subjects recognize special problems connected with complex adaptations, there are few general discussions on the topic in print; but the few are notable. The book by Rensch (1959) bursts with examples and insights relating to major adaptive systems. Mayr's book (1963), very likely the most complete compendium on evolutionary theory ever written, contains some of the most thorough considerations of the origin of major adaptations in print anywhere. Of equal importance, and a factor that makes Mayr's effort close to unique, is that his conclusions are tied to the generous regard for population biology presented between the same two book covers.

Admittance of population concepts to analyses of complex adaptations is a most crucial step. If the evolution of major adaptations seems to cause some mild discomfort, it is largely, if not entirely, due to an apparent difficulty in easily relating such matters to the level of the population. This is the level upon which evolutionary thinking has traditionally been tested. For this reason,

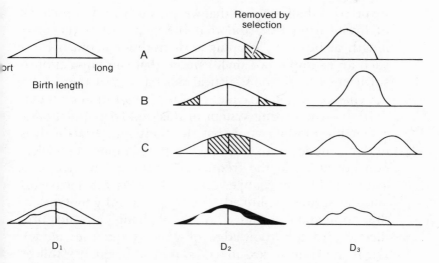

Frequency distribution of birth length in a hypothetical fish **5** population, showing effects of different modes of selection (in A–C) on birth lengths in the next generation. D. More realistic selection scheme, with selection not so intense and abrupt. D_1 compares curves just before and just after selection has occurred. D_2 shows the after-selection curve raised until it touches the larger curve at some point. D_3 shows the frequency distribution of the survivors of selection. Explanations are in the text. (D is from Van Valen, 1965a.)

and for better ones, the population is an appropriate proving ground for evolutionary hypothesizing of all sorts.

Traditional treatments of selection frequently proceed from noting statistical changes in a population due to the unequal survival and reproduction of different genotypes. Figure 5 shows a frequency distribution of some measurable character in an imaginary population. Since the situation is imaginary, we may propose anything as the character, and anything as the sort of animal involved. I shall suppose that we are studying a population of live-bearing fish, and that the character is the body length at birth. In keeping with traditional practice for such an example, we shall assume that the population is sufficiently large for statistical reliability, that there is no net change of gene frequency due to migration or mutation, that the mating system in this diploid population is essentially random, and that the body length at birth is under genetic control. In such a population, were selection not at work, the frequency of any given allele at a locus would not change, and so long as the locus was autosomal (not sex-linked), the frequency of genotypes in the population would also be unchanging (and predictable from a knowledge of gene frequencies alone). Or, if the locus *is* sex-linked, genotype frequency will be unchanging once the population has reached equilibrium. Under such conditions the population is said to be in a Hardy–Weinberg state [2]. True Hardy–Weinberg populations are rare, perhaps nonexistent, in nature. But they serve as idealized criteria for comparison with populations such as the one we are considering. Our fish population cannot be in a Hardy–Weinberg state because selection is changing the gene frequency from one generation to the next.

There are several fundamentally different ways in which selection can affect the distribution of birth lengths in the population. Let us suppose that selection favors a somewhat smaller birth length each generation, with the effect of moving the average birth length toward lower values, as shown in Figure 5A. This would be an instance of "directional selection," caused by selective removal of fishes carrying genes for a high birth length. This is not the only type of selection. In another possibility, shown in Figure 5B, selection favors the present average body length at birth but acts against individuals whose birth size is much different from the average. This type of selection, called "stabilizing selection," does not produce a directional shift of the mean but holds the mean to the same value while reducing the statistical variance. A third possibility, "disruptive selection," appears in Figure 5C, where selection chooses not one, but two, birth lengths to favor.

The "truncating" effect of selection shown in, for instance, Figure 5A is an oversimplification, although a useful one. Van Valen (1965a) has emphasized that truncation can be produced by artificial selection, but that in nature selection is usually less abrupt. He provides an easy method to demonstrate which part of the mortality in a population is selective. In Figure 5D, differential mortality has reduced the original curve to the smaller one, which then is drawn as reenlarged, inside the original curve, by raising all points of the smaller one proportionally until some part of it touches the larger curve. The point of contact is the optimum value of the character, and the difference between the shapes of the two curves shows the differential mortality due to selection. There is greater ease in thinking in terms of the simplified examples shown in Parts A–C of Figure 5, however, and in cases of strong selection the accuracy forfeited should not be unbearable.

Directional selection is easy to observe in the laboratory, and a number of investigations have been made in natural situations. For instance, the now-classic studies by Camin and Ehrlich (1958) on water snakes revealed a strong directional selection on color pattern. The sample populations were drawn from small islands in Lake Erie which, although lying out from the mainland, could often be reached from it by a swimming snake. The typically

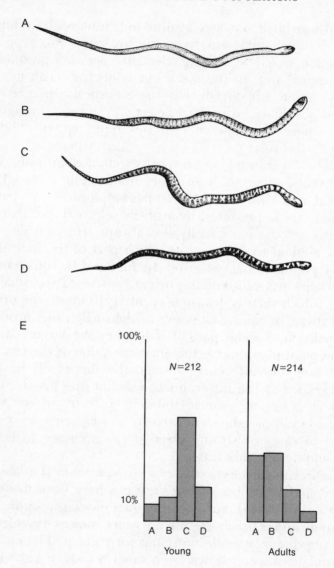

banded pattern of the mainland snakes was apparently disadvantageous on the islands, whose gray limestone shores showed up the banding all too blatantly to the keen eyes of predatory birds. A number of snakes in the island population showed a uniform gray with no banding whatsoever, and other individuals showed varying degrees of bandedness. This all points to the reduction in banding being favored by selection.

Camin and Ehrlich compared the banding patterns of island newborn snakes with those of island adults. The results are shown in Figure 6. Of the total range of patterns in newborn snakes, those individuals with less banding were more successful at surviving to adulthood. There is a directional shift in the average amount of banding from young to adult island snakes. If a barrier were built to restrict the occasional emigrant from the mainland, one could guess that the island snakes would all become bandless. But as long as "banded genes" continually find their way to the islands, this is quite unlikely to happen.

Examples of stabilizing selection are also found in nature. It is a common matter in snakes, for instance, to find that juveniles show a greater variability in scale patterns than do adults, owing to selective removal of the most extreme individuals (Dunn, 1942). A comparison of juvenile and adult foot scalation in several species of

6 Variation in color pattern in water snakes *(Natrix sipedon)* from islands in Lake Erie. Although the variation from unbanded to heavily banded is continuous, four basic "types" are indicated by letters A through D. E. The histograms compare frequency distributions of the four types as newborn snakes and as surviving adults (i.e., after selection) collected on one of the islands. Selection clearly favors unbanded types. (Redrawn from Camin and Ehrlich, 1958.)

Cuban lizards shows a similar stabilization (Collette, 1961).

Disruptive selection has been amply demonstrated in laboratory populations of *Drosophila* (Thoday, 1959). Occasionally it is argued that disruptive selection could produce such sharply distinct groupings within a population that two separate demes—genetically isolated breeding populations—could arise [3]. There is but meager enthusiasm for this idea, for most biologists are quite convinced that the evolutionary partitioning of genetically distinct demes, termed "speciation," must almost always require a geographic partitioning as a first step (Mayr, 1942).

These examples of selection have several factors in common. Selection removes certain individuals who possess certain genes, leaving an increased frequency of other genes in the population's survivors. When those survivors reproduce to form the next generation, the altered gene proportions are reflected in the gene frequencies of the offspring. Thus, returning to the example of our fish, if for some reason the individuals with shorter birth length survive better than others, more of them will reach the age of parenthood. As a consequence, the next generation will see a greater proportion of short newborn fish than existed a generation before.

This sort of selection changes gene frequencies by selective removal of certain potential parents, whose genes go with them. It is based on differential survival and must, of course, work only if there is differential mortality.

Not all selection requires mortality. Differential reproductive output does not depend on the demise of potential parents. In our fish example it could be possible that parents that produce smaller fish can thereby produce *more* fish. If smaller fish do not face a seriously reduced prospect of survival and successful reproduction, gene frequencies of parents producing the greater numbers will increase with each generation. The result will be that, even lacking selection on the survival of newborn fish of different sizes, the average birth size will decrease from one generation to the next.

It is probable that in most populations both kinds of selection [4] are at work (they might often, however, act to oppose one another). In all species imaginable there is a limit to population growth; and once the limit is reached, differential mortality will become a fact of life [5]. Both modes of selection are alike in that they reduce to the question of which genotypes can most increase their representation in future generations. This is measured by population geneticists in terms of *fitness*.

The effectiveness of selection in changing gene frequencies depends upon several factors. Mutation pressures that replace selectively disfavored genes will, quite clearly, work toward undoing the action of natural selection. The same neutralizing effect will be felt in cases of immigrations where disfavored genes are continuously carried in by the immigrants (which seems the case in Camin and Ehrlich's snakes).

Population size is itself a factor that can limit the effectiveness of selection. If the population is so small that it loses statistical reliability, accidents of sampling can change gene frequencies in ways that are unpredictable. In these cases the tendency of selection can be overwhelmed by statistical accidents. Gene frequencies in very small populations change erratically from one generation to the next, and eventually any allele at a given locus will become either lost or fixed at a frequency of 100 percent. This random fluctuation of genes toward loss or fixation is termed "genetic drift."

An obvious factor in determining selection's effectiveness is the intensity of selection. If the genotypes in a population were of very unequal fitness, the gene frequencies could be changed relatively fast. If, however, selective intensity were low because fitness differences were slight, the rate of change would be correspondingly small.

Perhaps it is less obvious that the gene frequencies themselves can determine the rate of change of gene proportions in a population. When selection is acting on a gene that is either rare or common, its effect is less than when the gene is at an intermediate level of abundance. Figure 7 shows some examples. When gene frequencies are extremely low, the number of individuals that carry the rare genes may be so few as to occasionally be eliminated through statistical sampling accidents.

In the long range, the effect of selection is more than the mere removal of some genes versus others. By increasing the frequencies of what once were rare genes, new genetic combinations can occur. Such new combinations stand at least a chance of producing new phenotypes [6] that selection can recognize and act upon. Furthermore, newly mutated genes will produce their phenotypic alterations in the context of the modified-by-selection gene pool of the population (Mayr, 1963). The very implications of mutations yet to come can be altered in advance by selection.

At each step selection acts primarily as a molding agent, making the best of whatever genetic alternatives lie before it. Without variability in the population, there is no selection. For example, migration of members from one population to another can bring new genes, which

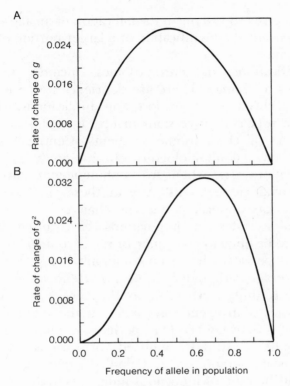

Rates of change of an allele's frequency (A) and a genotype's **7** frequency (B), depending upon the present frequency of the allele. The upper curve is for incomplete dominance, the lower for a completely recessive gene. In each case the gene has a 20 percent selective advantage ($s = 0.2$) when fully expressed. Note that the rate of change in either case is greatest at a frequency intermediate between zero and unity. (Modified from Falconer, 1960.)

instantly increases the genetic variation of the invaded population. A more fundamental mechanism is mutation, fundamental at least in that, ultimately, all genetic differences among organisms have arisen that way. However, "mutation" is a word used in different senses by population geneticists. It can mean the alteration of an allele at a single locus, or the modification of a larger portion of the chromosome.

Modifications of chromosomes include several categories of change. There are deficiencies, where portions of a chromosome are lost, and duplications, where some are repeated. Inversions involve a positional rearrangement of chromosome segments. Centric fusions combine two nonhomologous chromosomes as one. Translocations arise when two nonhomologous chromosomes exchange segments. Any of these chromosomal alterations can underlie phenotypic changes.

Mutation rates in populations have often been thought to be quite low. A range of rates from 10^{-4} to 10^{-6} mutations per locus has been commonly acceptable as a rough average, although it is not easy to be very certain about such things, especially as they occur in natural populations. Many geneticists feel that the vast bulk of mutations is detrimental. This, perhaps, is because harmless mutations can go unnoticed for various reasons. Recent reconsiderations (e.g., Ohno, 1970) are making it clear that the ratio of innocuous mutations to deleterious ones could be considerably higher than originally estimated.

The presumed rarity of new mutations has favored the suggestion that variation must usually depend upon mechanisms other than mutation. The most obvious possibility is that the genes already present in the population recombine in each new generation and provide a great array of new and varied phenotypes. In sexual reproduction each offspring will contain a genetic donation from each parent which could include novel—or at least unusual—gene combinations. Some of these combinations might be advantageous and thereby enhance the survival of individuals possessing them. But without some means of holding the combined genes together, the segregation of chromosomes in the formation of gametes will destroy the combination, and the selective advantage will be lost in the following generation of progeny.

This loss may very possibly be temporary, but it still might take considerable time to effect a restoration. If in each generation selection favors the combination whenever it occurs, the genes involved could increase in frequency as a result, ensuring the longer-range consequence that the occurrence of the combination will grow increasingly probable. This is a slow process and is, of course, not likely to happen should there be detrimental effects caused by the same genes when they are out of combination.

There are, however, mechanisms that can act to retain combinations. For instance, the chromosomal alterations of centric fusion and translocation combine different loci from nonhomologous chromosomes within a single, modified chromosome. But of much greater probability of occurrence is the crossing-over exchange of segments between homologous chromosomes. In fact, to many geneticists the term "recombination" refers exclusively to the collecting of the considered genes onto a single chromosome. As such, the recombining genes are linked together and segregate as a unit. For years it has been assumed that the crossing over possible in diploid species promoted new combinations of genes, including newly mutated genes. Quite recently, though, a detailed mathematical analysis has challenged the notion that the crossing-over mechanism enhances the linkage of new mutations on the same chromosome (Eshel and Feldman, 1970) [7].

The genetic events that contribute to the variability of

populations provide a slate of alternative phenotypes for the selection process. What I perceive to be the most widely held view is that the differences among phenotypes in an evolutionary progression are small. It is thought that organisms are quite in biological harmony with themselves and their external environments, and that any great or sudden alterations in phenotype will almost certainly disrupt this comfortable condition. Fisher (1958, p. 4) amplifies this notion by analogy with a microscope adjusted for distinct vision. The system may be deranged by slightly moving the lenses' position, changing their curvature, and so on, but "any large derangement will have a very small probability of improving the adjustment, while in the case of alterations much less than the smallest of those intentionally effected by the maker or the operator, the chance of improvement should be almost exactly one half." If this analogy is acceptable, it follows that if the individuals in a population are at an operational level of adaptation, large phenotypic variants will usually be selected out, leaving lesser, more easily tolerated variants behind. The implication from this is that evolution moves in small steps, and that adaptations are built slowly and gradually.

NOTES

1. De Beer (1958), Eden (1967), Goldschmidt (1940, 1955), Gould (1966), Mayr (1963), Rensch (1959), Sondhi (1962), Stebbins (1968, 1969), and Waddington (1957, 1962). See also Davis (1949), Frazzetta (1970), Olson and Miller (1958), Oxnard (1969), and Whyte (1965).

2. Any good basic text in population genetics will provide a theoretical treatment of Hardy–Weinberg populations, in-

cluding attainment of equilibrium in cases of sex-linked loci, and of alleles at several loci on different chromosomes. Books in this category include Crow and Kimura (1970), Falconer (1960), Li (1955), and Mettler and Gregg (1969).

3. See Thoday and Gibson (1970, 1971) and Scharloo (1971).

4. In considering these opposing selective forces it is always necessary to remember that it is not the number of new offspring produced that matters directly, but the number that survive to reproduce themselves. It is very possible to imagine adaptive equipment of a sort bestowing incredible survival properties on an individual when but a few individuals are produced by a set of parents—contrasted with an alternative, in which many offspring appear but none has the high probability of survival possessed by the lucky few in the first case. However, if the reproductive output in the second case were three times the first, a drop of one-half in individual survivorship would be a negligible price to pay.

5. This notion is embodied in the simplified logistic model of population growth, which shows the effect of population density on the growth rate:

$$\frac{dN}{dT} = r_0 N \frac{K - N}{K}, \quad K \geqslant N .$$

Here N is the population size at the time considered, K the carrying capacity of the environment for the species in question (it is a measure of how many individuals the environment can hold) and r_0 the maximum attainable rate of increase (see Chapter 3) for a population when it is under the special circumstances that population density factors that could limit population growth are at a minimum. Although there is no precisely defined relationship between r_0 and the *general* term r, which simply describes the population as it is when we observe it (see Chapter 3), the sense of the relationship is $r = r_0(K - N)/K$.

Clearly, in the logistic equation, as the number of individuals (N) approaches K, the rate of growth diminishes to zero (as the great size of the population presses upon critical environmental resources whose renewal rate cannot support further population growth). A graph of N against time (on the abscissa) shows an S-shaped curve, which indicates a slow, initial rate of change. At that period, although $r = r_0$ $(K - N)/K$ is high, the number of potential parents (proportional to N) is low. Later, when N is increased and r is still fairly high, the population increases rapidly until, later still, as $N \rightarrow K$, $r \rightarrow 0$, and growth effectively ceases when the population has attained the saturation number of individuals. General references on the model include Andrewartha and Birch (1954), Emlen (1973), Slobodkin (1961), and Wilson and Bossert (1971).

By no means should we expect that all natural populations are governed by limitations of resources needed to sustain individual life. Some populations are likely to be restricted by physical factors such as climate, or by predation, and so forth. In such cases the logistic model, at least in unmodified form, may not adequately describe factors that limit populations to a certain size level.

6. In some cases the extreme concentration of formerly rare genes in an individual genome can produce very novel and unusual characters (e.g., Sondhi, 1962).

7. Also see the discussion by Franklin and Lewontin (1970) on the stabilization of linked loci. Recently Watt (1972) has proposed a mechanism of mutation by intragenic recombination.

3

POPULATIONS
AND INDIVIDUALS

Evolution is a population phenomenon in that it is the change of populations in time. But adaptation is an individual phenomenon in that traits which do not aid the individual bearer have no selective advantage. The "adaptation of populations" is merely the observed effect, actually incidental, of the aggregate of adapted individuals. Selection acts on the individual through the differential production of offspring by certain individual phenotypes versus other individual phenotypes. If the phenotypic differences reflect genetic differences, selection will produce evolution. I view arguments that involve the "benefit to the species" as irrelevant. If such arguments ever become generally convincing, it is the signal that very many of our conceptual views about natural selection are wrong [1].

Nonetheless, individuals are members of populations, and the selective success and evolutionary future of their adaptations should always be considered in the context of the population to which they belong. This is not to say that an individual member of one species or deme (breeding population) does not interact with individuals of many other demes. It does, but selective factors from any source affect individuals in a way to produce changes in the size and/or composition of their population. For this reason we can get away with focusing on events within a single deme, even though selection due to interaction with members of another deme may be immensely apparent. There are several ways to get into the construction of a population growth model, and for most animals these are usually classified into two types. The type depends on whether we consider the production of offspring to occur during discrete time intervals—such as seen in species that breed once a year during a particular season (bullfrogs in Massachusetts, for example)—or continuous breeders where reproductive members of a species pay no attention to the season but are about as likely to breed today as seven months from now (humans, for example). In the case of discrete intervals we can readily make some calculations of future population size if we accept several assumptions.

Let us imagine that a population at the present time ($t = 0$) contains N_0 members. How big will it probably be next year? To calculate this we shall need information on the proportionate increase (or decrease) of the population from year to year. The proportionate increase will be expressed as r, and it is obvious at once that it must depend upon the average difference between births and deaths occurring per capita per year. In many, perhaps the majority, of natural populations, when viewed over large time intervals, the births and deaths more or less balance and give r a long-run average value of zero. I only wish that were true of human populations today, but it definitely is not. For any population, however, whether increasing or not, it is still pertinent to know about the factors that *tend* to increase it.

In our example, if N_0 is the population size at $t = 0$, the size at $t = 1$ is $N_1 = N_0 + rN_0$ or $N_1 = N_0(1 + r)$. If we assume that r remains the same over the years, the size a year later, when $t = 2$, is $N_2 = N_1 + rN_1$ or $N_2 = N_1(1 + r)$, or $N_2 = [N_0(1 + r)](1 + r) = N_0(1 + r)^2$. Still holding to our assumption of constant r, the population size after any time t is, in general,

$$N_0(1 + r)^t. \tag{1}$$

Much of basic population theory has been formulated not for the case of discrete breeding intervals but for continuously breeding species. A number of species are continuous breeders, but part of the motivation for using a continuous model stems from the greater ease in doing things that way. Discrete generations give rise to events that occur in sudden, discontinuous jumps which can be ponderous to express quantitatively. The continuous production of young involves smooth, uninterrupted phenomena that are more manageable even if they do require somewhat more sophisticated mathematics. We shall take the easiest path and consider the case of the continuous model. I shall want to start along this path from the place of our last quantitative statement, formula (1).

In order to proceed in the direction of continuous births it is recognized that we must consider very many reproductive periods in a year, not merely one such period. Formula (1) is in the same form as a compound-interest formulation for money earned in a savings account at the bank. By that analogy, the value given by the formula reveals how much money has accumulated after t years, while N_0 is the original amount invested and r is the interest rate. If the yearly interest were compounded n times in a year (instead of only once), the actual interest figured at each such period would be r/n, and since there would be n such periods in each year, formula (1) becomes

$$N_0(1 + \frac{r}{n})^{(n)t}. \tag{2}$$

Convenience is gained if we set $r/n = 1/u$, ($1/u$ simply being a term plucked from midair but whose utility will become apparent momentarily), so that $n = ur$. Now formula (2) can be written

$$N_0(1 + \frac{1}{u})^{urt}$$

or

$$N_0[(1 + \frac{1}{u})^u]^{rt}. \tag{3}$$

The actual number of times in the year that interest is compounded has not yet been specified. But if, in fact, n becomes indefinitely large, so that there are an infinite number of times when interest is compounded, the situation becomes one of continuous compounded interest—or of continuous births in the population. As n approaches infinity ($n \to \infty$), so does u ($u \to \infty$) while ($1/u$) $\to 0$, since dividing any finite number by an immeasurably large one results in an immeasurably small value, one that approximates zero.

After allowing n to become infinitely large, the term $(1 + 1/u)^u$ in formula (3) becomes especially interesting. This is so because although the value of terms within the parenthesis is only larger than unity by an immeasurably small amount, that value is raised to an infinitely high power. The overall result, as can be proved mathematically, is that as n approaches infinity the expression $(1 + 1/u)^u$ tends toward the definite finite value of $2.71828...$, which is always designated as e and forms the base of the system of natural logarithms.

Formula (3) can be modified to fit the continuous case, and can be equated to N_t, the population size at time t, when u is allowed to approach infinity (designated $\lim_{u \to \infty}$):

$$\lim_{u \to \infty} N_t = N_t = N_0 e^{rt}. \tag{4}$$

This expression is flexible enough to look backward in time as well as forward. As written, (4) gives the population size t years from the time when $t = 0$. Instead we could ask how big the population was t years *before* the time $t = 0$, in which case we want to know N_{-t}, which is $N_0 e^{r(-t)} = N_0 e^{-rt}$.

From this point it is easy to proceed to a related matter—determination of the contribution of each element of the population to the maintenance of the present state of population growth (which need not be positive, but is zero if the population is not changing and negative if it is declining). The problem can be handled most easily if we assume that r is constant and that it has remained constant long enough to permit the population to reach a state of equilibrium. A population in equilibrium is not necessarily unchanging in size (unless $r = 0$), but the age distribution of individuals has attained stability. A stable age distribution is assurance that the proportion P_t of individuals whose age falls within the interval of ages from t to $(t + \Delta t)$ will remain the same from one generation to the next. This is true even if the total *numbers* of the P_t members are changing, for, after all, under these conditions the total of all others in the population is changing in the same manner [2].

To get at the matter of the relative contribution of each element in the population, we must consider a few more factors. Mentally we could sort through the entire population and classify each member according to his age. This classification will require the establishment of some agreeable system of age intervals. For example, it could be ambiguous to speak of an individual aged two years, for it might not be clear if I mean to include everyone aged at least two as long as they have not reached their third birthday, or if I am rejecting from that category anyone older than two years and five minutes. The interval we choose might be a matter of years, weeks, seconds—or even less—and can be expressed in general mathematical form as Δt. By this convention population members can be classified by age according to the intervals into which they fit, so that when we speak of an individual of age t, we really mean that his age lies somewhere in the interval t to $(t + \Delta t)$.

We have been talking about Δt as if it were a finite, therefore measurable quantity. But a population model based on continuous production of offspring should avoid discrete "steps" or breaks. Something must, then, be done about Δt, since its finiteness breaks the spectrum of possible ages into a number of discrete steps or categories. Of course, this will be the actual situation in any real population since the population itself is of finite size. But if the population is large, we could pretend, without too much penalty, that the ages represented range from zero to the most venerable possible in a continuous fashion. It is easy to see that as we reduce the interval Δt to a smaller quantity, thereby increasing the number of intervals within any unit of time, we shall come closer to a continuous array of ages in the population. When Δt is diminished to an infinitesimal, mathematical continuity becomes realized. In the convention of calculus we simply replace the Δ or delta symbol with the letter d to remind us that our interval—now written dt—is immeasurably small.

It will now become clear that if b_t is the probability that an individual of a given age t will produce offspring in the time interval t to $(t + dt)$, his proportional contribution of births occurring in that time will be $b_t\,dt$. This term for the birth of offspring to a single individual aged t can be compared to a birth factor for the entire population. We shall let Y represent the number of young born per capita of the entire population in an interval dt. Thus if the population size t years ago was $N_0 e^{-rt}$, then $N_0 Y e^{-rt}$ is the total number of births that occurred exactly t years ago.

Suppose now that the chance of an individual surviving t years is l_t. We then note that $N_0 Y l_t e^{-rt}$ is the number of individuals born t years ago who are alive at this moment. To put it another way, the expression gives us the number of individuals who are now aged t years.

It is easy to see that the number of births now being contributed by this age group is $N_0 Y e^{-rt} l_t b_t \, dt$. And from this we can readily arrive at a formulation for the total births contributed at the present time by the entire population, that is, the present contribution of all the age groups. To do this we simply add together the contributions of each age group. We will start with persons aged zero ($t = 0$) and continue to the impossibly old age of infinity ($t = \infty$). Neither of these extremes can, of course, contribute offspring, because individuals aged zero are just being born themselves, and no individual lives to infinity. But between them is a span of ages during which offspring can appear, and our summation over all ages will surely regard all possible births. Our summation, then, could look as follows:

$$N_0 Y \left\{ \left[e^{-r(0)} l_0 b_0 \; dt \right] + \left[e^{-r(dt)} l_{dt} b_{dt} \; dt \right] + \left[e^{-r(2dt)} l_{2dt} b_{2dt} \; dt \right] + \right.$$
$$\left. \cdots + \left[e^{-r(\infty)} l_\infty b_\infty \; dt \right] \right\}$$

where we allow t to be 0, then dt, and so forth, building its value in infinitesimal steps until it finally arrives at $t = \infty$. Gratefully there is a shorthand method of writing all this by using the integral ($\int$) notation of calculus:

$$N_0 Y \int_0^\infty e^{-rt} l_t b_t \, dt. \tag{5}$$

The integral sign simply indicates that we are adding up a large series of infinitesimal quantities. The symbols 0 and ∞ at the bottom and top of the integral sign denote that we are going to let the variable t begin at zero and assume successive increments in "steps" of amount dt. During each such step the term inside (i.e., to the right of) the integral sign assumes an infinitely small numerical value which is added to the values summed from all the preceding steps. This goes on until t reaches ∞ and the addition process ceases. The term $N_0 Y$ is placed outside (i.e., in front of) the integral, because it is constant and its value does not depend on the value of t. It is thus different from such terms as l_t, whose value very much depends on t; we can, for instance, be reasonably certain that l_1 is greater than l_2.

Our last equation (5) gives us the total births produced by all population members at this moment. A simpler way to express the very same thing is by using the term $N_0 Y$. Notice that we have proved what was intuitively obvious all along—that

$$\int_0^\infty e^{-rt} l_t b_t \, dt = 1 \tag{6}$$

which is to say that the proportional contribution to the present births of all members together is unity.

Obvious as this relation is, it is useful to us when we consider a more important matter, the relative importance of each individual of a given age group to the maintenance of the present state of the population. Clearly not all ages are equivalent in reproductive potential. How does their "value" to the population change through life? Sir Ronald Fisher posed this question—and its mathematical solution—basing much of his thought on the considerations we have just discussed, which, incidentally, are largely the work of A. J. Lotka [3].

We shall represent the "reproductive value" of each individual aged x years by v_x, and compare this to the value (v_0) of a newborn individual. Hence we seek an expression equivalent to the ratio v_x/v_0. Let us deal first with v_x. Since we care about the *average worth of each individual* of age x, we can be sure that we will have to divide the total value of all x-aged members by the number of such members. Accordingly, the first term on the right side of formula (7) is $1/N_0Ye^{-rx}l_x$. The long, complex term in brackets is the sum of relative contributions, present and future, of offspring by all individuals who are presently aged x. Each numerator is the number of young contributed at each age, starting with the present age of the x-aged individuals and progressing—from numerator to numerator—to ever-advancing ages. The denominators provide the number of young produced by the total population at each interval. This gives us, for each set of numerators and denominators within the large brackets, the contribution of persons now aged x relative to that of the whole population at the same time. This mathematical leviathan,

$$v_x = \frac{1}{N_0Ye^{-rx}l_x} \left[\frac{N_0Ye^{-rx}l_xb_x\,dt}{N_0Ye^{r(0)}} + \frac{N_0Ye^{-rx}l_{x\,+\,dt}b_{x\,+\,dt}\,dt}{N_0Ye^{r(dt)}} + \cdots \right.$$

$$\left. + \frac{N_0Ye^{-rx}l_\infty b_\infty\,dt}{N_0Ye^{r(\infty)}} \right] \tag{7}$$

can be simplified algebraically to

$$v_x = \frac{e^{rx}}{N_0 Y l_x} \left[e^{-r(x+0)} l_x b_x \, dt + e^{-r(x+dt)} l_{x+dt} b_{x+dt} \, dt + \cdots \right.$$

$$\left. + \, e^{-r(\infty)} l_\infty b_\infty \, dt \right] = \frac{e^{rx}}{N_0 Y l_x} \int_x^\infty e^{-rt} l_t b_t \, dt, \qquad t \geq x. \qquad (8)$$

Since v_0 is the special case of v_x where $x = 0$, modest reasoning takes us to

$$v_0 = \frac{1}{N_0 Y} \int_0^\infty e^{-rt} l_t b_t \, dt$$

which, from equation (6), becomes

$$v_0 = \frac{1}{N_0 Y}. \qquad (9)$$

The reproductive value v_x / v_0 is simply the ratio of equations (8) and (9), which is

$$\frac{v_x}{v_0} = \frac{e^{rx}}{l_x} \int_x^\infty e^{-rt} l_t b_t \, dt. \qquad (10)$$

Actual data that allow the calculation of reproductive values are available for a number of animal species, including our own. R. A. Fisher performed such calculations for a human population whose positive r value ($r = 0.0123$) shows that this population was gaining in numbers. The curve shown in Figure 8 is shaped rather like those drawn for other animal species. It rises to a peak value rather early in life—in this case at approximately eighteen years of age—and thence declines quite rapidly, vanishing on the abscissa at the point corresponding to the oldest possible age of reproduction.

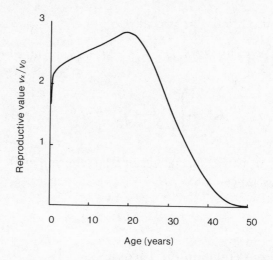

8 Reproductive value of a human population. The calculations are based upon female population members, a simplification generally employed by demographers. (From Fisher, 1958.)

In reproductive-value curves the peak generally occurs just at, or somewhat after, the birth of the first offspring, depending on whether a set of parents produce offspring on one or on several occasions. The curve's initial ascent toward the peak is due partially to the term e^{rx}/l_x in equation (10), which is the reciprocal of the proportion of individuals living to age x. Fewer individuals survive to progressively older ages that approach reproductive maturity. Those who do are relatively more valuable, as the role of maintaining the population falls to them. Moreover, in a growing population, more individuals are to be born in the future than were born in the past. The number of parents are, then, always fewer than the surviving offspring they must bear. Members of older age groups were born when relatively few births were occurring; hence individuals of older ages are comparatively rarer than younger ones and toward the onset of reproduction are relatively more valuable.

The descent of the curve at ages beyond the reproductive peak is due to the relatively greater value to the population of the parents' earlier offspring and the increasingly lowered prospects for future survival (l_x) and for reproduction (b_x) of an individual. The curves shown in Figure 9 compare several vital factors involved with Fisher's reproductive-value concept.

The reproductive curve is intended to show the average individual's reproductive value to the population throughout his life. But when practiced on species in which individual parents care for their offspring after birth, the method has a major flaw. It disregards the value of parents to young already born. In our species, this consideration, if entered into the calculations, would delay the peak by several years.

Reproductive value is a population concept. It

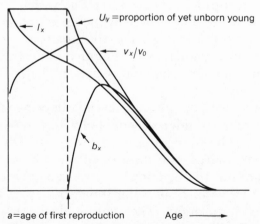

U_y =proportion of yet unborn young

l_x

v_x/v_0

b_x

a=age of first reproduction   Age

Intuitive comparison of several population parameters. The **9** vertical scale is interpreted in terms appropriate to each curve. In these terms, the initial value of each curve is $l_0= 1$, $v_0/v_0 = 1$, $b_0 = b_{a-dt} = 0$, and $u_{y0} = u_{y(a-dt)} = 1$.

mathematically describes the worth of an individual to the population but not to its own genotype. For that reason it is a mistake to interpret the reproductive-value curve drawn for a population as an indicator of changing selective intensity throughout an individual's life. As viewed from the individual's standpoint, the significance of his own survival does not await an age when he becomes important to the population. Indeed, to attain any age he must survive during all earlier ages. It is thus important to recognize the distinction between the individual's value with respect to his own genotype—which has a selective basis—and his value with respect to a group.

Groups and individual genotypes need not, however, be regarded as inalienably separate, at least in all definitions of them. Parents and their offspring form a group in which many identical genes are liberally shared. An intuitive bending of the mathematical concepts that we have been pondering permits us to apply them to such genotype groups and thence to envision what changes, if any, in selective intensity may be experienced by the individual. Alterations of selective intensity are highly significant in the analysis of adaptive patterns. The ontogeny of complex adaptations may well depend upon such changes.

To begin with, let us imagine the offspring of a single set of parents considered separately from others in the population. The young animals grow to maturity, become parents themselves, and have offspring who continue in similar fashion. In our mind's eye we can envision such a cohort of individuals as being somewhat apart from the remaining population. The genetic distinctness of the cohort from the population at large will gradually dissolve with passing generations since the cohort's members are free to mate with "outsiders" and parentage is shared.

But to a limited degree we may think of an "r" value for the rate of increase of the cohort as being somewhat distinct from the population r. If the cohort's "r" is greater than that of the overall population, the genes characterizing the cohort will increase in frequency. The evolutionary result of this is that the population will tend genetically toward the cohort. The converse, when the cohort's "r" is lower than the population's, is of course true also. Selective favor—or disfavor—is reflected in the "r" value of the genetically similar members of such a cohort.

Selection will favor those parents who can contribute their genotypes to future generations. This can of course be done only via the parents' offspring. Then from the selective standpoint the young ones increase in value to their parents from the time of birth to some period at or shortly following the onset of reproductive age. Thereafter the reproductive value (defined now in terms of the parents, not of the entire population) of these offspring will decline. This restricted interpretation of reproductive value, applied to a single set of parents may, with appropriate cautions, be closely compared with the broader, population-based concept. But whereas the population concept has little direct bearing on individual adaptation, the restricted concept does. Those offspring who are approaching the peak of the curve are more important to the survival of the parental genes than younger progeny. From the standpoint of optimum strategy, parents faced with the task of caring for offspring of various ages will often incline toward a greater expenditure of effort on those nearest the reproductive value peak. This favoritism will be expected only when older offspring derive sufficient benefits (compared to benefits for younger offspring) from a given amount of parental effort.

In few species do parents provide postnatal care with the elaborateness of humans, or even of house sparrows. But parents of all species donate some effort, therefore care, to the development of their progeny. This effort may be no more than the carrying of a developing embryo by a lizard, or the production of an egg—to become an embryo only after its release and subsequent fertilization—by a teleost fish. In some cases the parental care is minimal, often it is moderate, not rarely it is considerable; but it is ubiquitous.

Although familiar and obvious, it is a significant paradox that as parents donate efforts to their offspring as the biological means to their genetic representation in future generations, the exertion of those efforts takes time and adds precariousness to the parents' own survival and, hence, to their opportunities to create future offspring. The longer the period of parental care, the greater is the risk to the parent. Although it seems severe to say it, the truth redeems the assertion that if the offspring are not to survive, they had best die soon.

Adaptation exists on both sides of an individual's birthday; embryos, like young and like adults, are adapted to their circumstances. During ontogeny, adaptations change in character in sufficiently complicated fashion that it is unlikely that all ontogenetic stages can be equally well equipped for survival. If inherent weaknesses must show themselves, the worst time for their manifestation is toward the end of parental care. If the time of emergence of such weaknesses can be governed by natural selection, the weaknesses will arise as early as possible; the period of greatest risk to the offspring will soon be got over with. In that manner the parents can be given the best chance of survival to reproduce again, and although the individual offspring may be lost, the parental genes may yet have a future.

There is, then, reason to expect that there will be a tendency for progeny nearer the end of parental care to be better adapted than younger progeny. When this is not true, it probably reflects a lack of complete genotypic/phenotypic flexibility in permitting the most detrimental weaknesses to be scheduled for earlier ages. Moreover, an accumulation of *all* weaknesses at any particular early stage would likely ensure disaster for all progeny, and selection would surely favor some spacing out of potentially harmful factors. Where many weaknesses are present, they might be most advantageously strung out over a long period rather than being confined to a narrow age span in early life, so as to reduce successive overlap of too many inadaptive conditions. Nevertheless, such spacing will probably not be distributed linearly with time, with the intervals between weaknesses being of equal duration.

For one thing, the ontogeny of an individual is not a smooth, even process lacking sudden events and accelerated transitions. For example, in its beginnings an individual mammal must be successfully implanted in the uterus—a sudden but critical change in its progress. It must successfully put to work its newly developing organ systems, which take over new, vital functions. Birth itself, and the physiological and behavioral changes that follow, are further discontinuities in the developmental schedule [4]. If the timing of adaptive strengths and weaknesses can be controlled through selection, there is at least a subtle expectation that at those necessarily critical times in ontogeny when the animal suffers high risk of mortality, other inadaptive influences are withheld [5]. If true, the scheduling of these major, risky events influences the overall adaptive picture, making a smooth spacing of adaptive weaknesses unlikely.

On the other hand, the advantage to the parent of

early mortality of offspring (when there is to be mortality at all) will tend to concentrate weaknesses toward earlier ontogenetic stages. If development did not necessarily involve discontinuous steps associated with periods of high risk, the spacing of weaknesses might fit an expectation of a reversed logarithmic distribution with a greater crowding of inadaptive aspects, which depress survival chances, in the earliest ages, and a greater reduction in frequency of such phases as time passes. There is no reason to think that selection does not account for all these factors—the necessity of spacing, the presence of necessary but dangerous and critical periods in development, the advantage of early versus late mortality of young—in a compromising fashion. Hence a tendency by selection to distribute the probability of mortality along some reversed logarithmic scale might be frustrated in many cases by the genetic inflexibility of the timing of critical events in the life history. Nevertheless, if this is the tendency of selection, even if its accomplishment in this direction is not total, neither should be its failure. Levels of animal adaptation should often be dependent upon ontogenetic age, and consideration of selective factors might teach us how to look at them.

But, in fact, can selection ever influence the time of occurrence of inadaptive periods in an individual's lifetime? Indirect evidence suggests that it can. And interestingly enough, much of this evidence comes from a consideration of the reproductive-value curve on the right side of the peak.

Just beyond the peak, as the eye moves left to right, the sharp decline seen there is in part due to the lowered chances of survival and reproduction. This condition, known as "senescence," is so familiar that it usually goes unquestioned, if not unnoticed. P. B. Medawar (1957) makes the point that organic parts are not metallic machinery—they should not wear out. Yet they seem to, despite the fact that repair and renewal mechanisms are effectively in evidence in early years. But the repair process falters as age advances, and the biological potentials of survival and reproduction are diminished.

In the same year, 1957, Medawar and G. C. Williams each published a mind-prodding interpretation of senescence; in their general approach, the works had a high level of agreement [6]. Their argument may be represented by the following example, which I have modified only slightly from Medawar's comfortable exposition of the idea. The basis of the example is the investigation of individual survival in a population of things that show no senescence whatsoever. In this case, then, an individual of any age has the same chance of survival to the next age category as any other individual, regardless of his age. There is no tendency toward greater frailty, susceptibility, or vulnerability of older members as contrasted with their younger associates in the population.

Populations of this kind are perhaps nonexistent in sexually reproducing animals. For illustrative purposes, we may have to invent our own population, probably an inorganic one, to purify the point. Let us imagine a very large population of teacups in a busy restaurant. The cups are used so frequently that they suffer a 10 percent chance of breakage in a month's usage. Accidental breakage (mortality) is not, let's assume, dependent on age in any way; a new cup is as susceptible to elimination as an old cup. (This assumption, incidentally, might actually be false in the real world of teacups.)

Unless the cups were replaced at frequent intervals, the restaurant would soon be cupless. So each month the proprietor buys $0.1 \times N$ new cups, N being the total

number of cups he is trying to retain intact in his establishment. This purchase plan replaces the damaged cups
and represents, as far as the analogy with a population of
living organisms is concerned, reproduction of new
members at a rate that, on the average, exactly makes up
for the loss due to mortality. I wish to pretend further
that any batch of cups acquired in any month bears a
slightly different pattern, so as to distinguish it from any
other batch. In this way the proprietor, aided by an
incredible memory, can look at any cup and know when it
was purchased. In terms of our analogy with living populations, then, the age of any individual can be assessed
immediately.

The last assumption to be made is that the proprietor
at some point in this process of losing and replacing cups

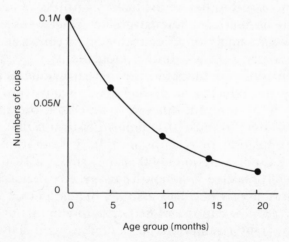

10 Survivorship curves for teacups with 10 percent "mortality." It
shows the number of cups now existing in each age group (and
thus indicates how many of an original batch of 0.1N will
survive to advancing ages).

records the number of cups in each age batch which he has on his shelves. Were the data from such a record presented as a graph, something like Figure 10, a survivorship curve for individual cups could result. It tells of the actual survival rates of cups at different ages—as we have defined ages for cups. But it tells something more important: the chance that any cup just now appearing in the population (being born, by analogy) has of living to a given future age.

No more than a glance at the curve reveals that the chance of surviving to an advanced age is decidedly worse than the chance of merely surviving for a few months. Yet, as we have defined things, the cups do not age. The survivorship curve depicts nothing more than the effects on survival of "natural" accidents that can occur to any individual regardless of his chronological seniority. The significance of this curve for any individual just entering the population is that his chance for survival to an age nearing fifteen months clings to the bleak side of hope. In a population of real living things, the survival chances of an individual, even disregarding all mortality "due to" senescence, diminish noticeably with age. In some species the reduction in prospects of survival may be noticeable over months, as in the teacup example. In others it may be reckoned in days, in still others, in decades; but all in all, a newborn individual can look forward to greatly worsening survival prospects as age advances, even when senescence is not a factor. This basic fact of life has undoubtedly influenced the manner of operation of selection.

The effect of senescence on the survivorship curve is to make even more unlikely the continuing of life beyond a certain age. Clearly the fact that senescence exists attests to the presence, in the individual's genome, of hereditary factors that work against his survival with increasing force as the animal ages. That natural selection has not weeded out these deleterious factors suggests that it cannot do so. One must suppose, then, that removal of those genes that promote senescence would have an overall effect of greater harm for lesser benefit.

Presumably many heritable, advantageous modifications which result from gene mutations or recombinations can involve relatively minor undesirable effects. This dualism in expression of genes may be viewed in the classic sense of gene pleiotropy. Or with less formalism it can be acknowledged that the perfection of one biological system will most surely afflict some others. It may be hard to modify one aspect of the animal to advantage without having to pay some price for it. Of course, the cost cannot be too high, relative to the advantage, or the new modification will not receive selective favor. However, the cost can be reduced substantially if the payment deadline is postponed.

Imagine a population in which senescence is yet unknown, whose survivorship curve for individual life expectancy is like that for our teacups. A new, heritable modification appears which may or may not be favored by selection. There are two major effects of the change; A, which is favorable, and B, which is unfavorable. The effects are separated in time as I have shown in Figure 11, from which it is not hard to appreciate the relative importance of A's time of occurrence. Effect B occurs significantly later. Even if B is as disabling as A is beneficial, its harmful influence is reduced in importance since it is not felt until the individual's prospect for survival has already dropped. With sufficient separation of B from A, the animal may not survive to the age when B is expressed, and so the potential harm of the modification may

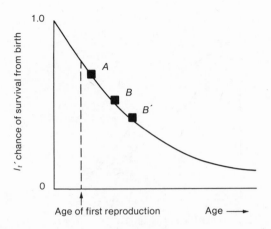

Schedule of appearance of a beneficial trait *(A)* and associated **11** harmful trait *(B* or *B')*.

have no actual effect on him. In such a case selection would favor the new modification, which, then, would eventually be shared by all individuals in the population.

More than that, if it were possible to shift effect *B* into older age—to position *B'* say—selection would certainly favor the shift. The further the bad effect can be pushed into older age, where survival chances are getting dimmer, the less will be the real injury due to the potentially harmful effect. The delaying of a few bad effects until later life will facilitate, by increasing selective advantages, the delay of other bad effects. Once started, the process becomes, as Medawar notes, self-enhancing.

With selection accepting mutations with ill effects in later life, and with selection pushing existing inadaptive expressions as far along as possible, an animal accumulates its best characteristics relatively early in life and its worst later on. Since many, perhaps most, adaptations have both good and bad aspects, a tremendous store of unhappy effects await the individual in late life to affect

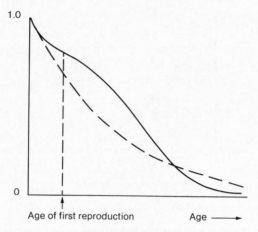

12 Comparison of survivorship curves. The dashed curve shows survivorship with a constant chance of mortality. The solid curve shows survivorship when mortality probability is lowered in early years but increased (senescence) in later years.

many parts of his being, not just a single feature. As Williams concludes, senescence is a "generalized deterioration, and never due largely to changes in a single system."

The consequences of selecting for early expression of good effects, and accepting later expression of bad ones, are rather as expected. Where good effects are concentrated, survivorship is relatively better and is shown by an upward warping of the survivorship curve in this region. But where senescent effects are concentrated, the curve is warped downward. These contortions of the curve are shown in Figure 12. As the solid curve shows, following a period of high viability there comes a rapid decline as a result of senescence. Animals in nature seldom experience this portion of the curve—they simply do not live that long. Expressed senescence is, in fact, an artifact of a protective society in most cases, and although rare in wild things, it is not so in zoo animals, house pets, and human beings, who receive those benefits of modern civilization that reduce the risk of accidental death.

Of course, the major point of the foregoing thoughts is that adaptation is not expected to be constant. The ontogenetic formation of complex adaptations must of necessity be a complex affair, presumably all the more so due to the exigencies of scheduling the best and worst effects of any adaptation. A proper study of animal adaptation should take these factors into account, search for evidence of them, explain their absence, if absent, and note their overall bearing on the animal's life and form if present.

If we can accept, in theory at least (and most is unspoiled conjecture) the view I have tried to develop regarding the timing of adaptive "strengths," we arrive instantly at the research of Lamont Cole (1954). Cole made a series of calculations for some hypothetical species of animals, for each taking into account the average litter size, the age at which reproduction begins, and the value of r achieved. Figure 13 is a slight modification of Cole's graph showing, for species that reproduce every year following the first reproductive year, the way in which r varies with the time of first reproduction. A glance at the figure is enough to see that early reproduction tends to be graciously favored by selection, so much so that following Cole's method, it is better to produce only one individual per litter if reproduction begins at age one year than to produce a litter of ten starting in the sixth year. The r value of about 0.63 in the first case is slightly higher than in the second, and very much higher when the producer of ten per litter withholds reproduction beyond the eighth or ninth year.

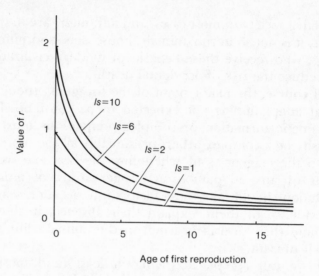

Age of first reproduction

13 Effect of early onset of reproduction on the value of r. Four kinds of hypothetical animals are shown, each of which has a different expected litter size, ls. (Modified from Cole, 1954.)

The effect of early reproduction is to shorten the generation time so that intervals between the production of children, grandchildren, and great grandchildren are diminished. In a single population the cohort of genetically related individuals with the shortest generation time will have an "r" higher than the overall population r, and are thereby being selected for. All other things being equal, of course—and therein lies a complication. If all other things were *always* equal, selection would favor a reduction in development time of the individual, with the result that large animals, complex animals, and animals with intricate learned behaviors would all be losers. The fact that you and I are both here speaks loudly that the "other things" are not, in fact, always equal.

Perhaps the most major "other thing" that is not

always equal is the death rate. Cole ignored the death rate in his calculation of r values, a perfectly reasonable approach since his only concern was with births. Since r is the difference between birth and death rates, its value can be boosted by a higher birth rate (which Cole demonstrated would result by shortening generation time) or by a lower death rate. There must be many animals in which the forestalling of first reproduction provides more time to an individual for further development, physical and/or mental. And in these there must be some benefit in reducing the death rate. Hence there are times when the benefits of delayed reproduction outweigh the disadvantages [7]. This realization does not, however, bring us to a shrugging off of the implications of Cole's work. Instead, the work provides an intuitive realization of the cost of complex development and how very great the benefits of delayed reproduction have to be before selection can grant approval [8].

The hypothetical advantage of early reproduction has an additional implication. The shape of any one of the curves in Figure 13 is roughly similar to the survivorship curve that can be drawn for teacups. In the case of the survivorship curves, it was clear why selection would place a particularly high premium on the early ages of an individual. The reproductive curves suggest the same thing. For both the reasons of survivorship and of age-related reproductive advantages, selection may sacrifice adaptive strength in advanced age for vitality in earlier years.

In Figure 14 I have attempted to summarize, in a wholly imaginary manner, the ways in which the several factors regarded in this chapter might occur in relation to one another. There was no conscious intention to describe, in this fashion, animals of a particular sort. Need-

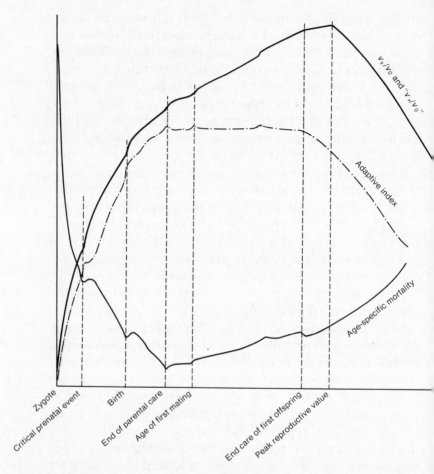

14 Comparison of several curves pertinent to individual adaptation. Explanation is in the text.

less to say, if one had a particular animal in mind, the figure would have to be modified accordingly.

The abcissa of the graph sets out a schedule of events that an individual of some imaginary species can expect to encounter. His life is considered to begin when he is a zygote, although one could successfully argue that it really begins earlier, and continues to an encounter with a deliberately unnamed critical phase during embryonic existence. Birth is regarded as another critical and potentially dangerous event, as is the termination of parental care, when the animal must manage utterly by itself.

The ordinate axis represents several scales, unspecified so that we may be more daring in drawing curves. The curves depicted are not compared with respect to actual values of points projected on the ordinate, but are compared only in a qualitative sense: What is their general shape, where does a curve change direction, and in which direction?

The age-specific mortality curve shows the chances that an animal has of *not* surviving from any given age to the next age interval. The curve is drawn to show increased chance of mortality in early and late life, a familiar pattern in several species, including our own. It also is drawn to show my assumptions that mortality increases during the critical periods.

A reproductive-value curve for the population as a whole (v_x/v_0) is drawn to include prenatal existence. I have assumed that any cohort in the population that might draw our interest is rather average in all respects, and an attempt to estimate its reproductive value might result in a curve quite like that of the population at large. For the reason of this expectation, the cohort's reproductive curve ("v_x/v_0") is superimposed on that which describes the population.

The remaining curve I have called the "adaptive index." The notion of the adaptive index is similar in spirit, if not method, to a concept developed by Emlen (1970) in an analysis of age-related ecological factors. The adaptive index is intended to represent the effect on the individual of the scheduling of adaptive weaknesses that I hypothesize. If the presence of such weaknesses cannot be altered by selection but their time of expression can, the adaptive index suggests how the total adaptive level of an individual—the "goodness" of his adaptations—may change with age. From acceptance of the arguments already offered, the theoretical optimum adaptive level should occur at the age when parental care ceases and last at least until the age of first reproduction. Prior to that time the selectively best adjustment of weaknesses will depress the adaptive index with increasing severity in earlier years.

Accordingly, the adaptive index is roughly shaped as a somewhat logarithmic curve which levels out at the end of parental care. After the individual has reproduced and has had sufficient time to care for his own first offspring, the levelness gives way to a decline. The actual shape of the curve may involve a few bumps, which correspond to critical periods in the individual's life. They would represent rises in the adaptive index to compensate for the higher risks at those times.

The adaptive-index curve roughly parallels the reproductive value to the end of parental care. This is not unreasonable. Nor is it unexpected that in its shape it could, in many instances, reflect a sort of mirror image of the age-specific mortality curve. There are several factors which relate the adaptive index to the other two curves, and indeed, the reproductive value and mortality curves are connected [9], for as Fisher said, "It is probably not without significance . . . that the death rate in Man takes a course generally inverse to the curve of reproductive value."

Of course the initially low level of the adaptive index is not to imply that during its earliest stages the organism is not a beautifully fashioned specimen of adaptive machinery. But there are implications that doubtless are real, which encumber speculation on whatever validity the adaptive-index concept might have.

As an example, the adaptive-index curve shown in Figure 14 does not take account of the discontinuity between adaptive weaknesses: the fact that weaknesses that might accrue to a blastula will have little to do with those involving a late gastrula, and less to do with an embryo that is nearly full term. Such matters could be "designed into" the curve, but this and other considerations would, at this juncture, be theoretical refinements of an ethereality.

As another example, one should not be misled to expect that periods of high mortality are always accompanied by a low adaptive-index value. There may be critical periods, sometimes of long duration, which subject an animal to extreme risks or hardships. During such spells the adaptive index might, in response to selection, maintain a tremendous value. For instance, a one-day-old fish can, because of its small size (if for no other reason), fall prey to many more predators than an adult. An age-specific mortality curve would reveal a fearfully low chance for survival of the baby fish. The disadvantage in being tiny is considerable, but tinyness itself is inescapable. During babyhood, then, the harshness of the fish's existence may require adaptations of excellent workability if there is to be any chance at all for survival. It is even possible that at no other time in life has the fish a more impressive adaptive apparatus, despite the high mortality at this age.

In the following chapters the relationships between individual adaptation and population factors must always be a minor theme, playing unceasingly in the mind's ear. It has never been correct, and never will be so, to approach the evolution of adaptations without placing the process in the context of the population in which it is all happening [10].

NOTES

1. The concept of individuals undergoing selection to aid the population to which they belong is termed "group selection." We shall deal later with "aid" exchanged between members of families, among whom many genes are shared. But where genes are not widely shared, as between an individual and the population at large, there is no selective mechanism favoring an altruistic individual over one who is indifferent. Apparent cases of altruism all seem, when studied carefully, to benefit directly the "altruist" or his genetic kin (see Williams, 1966). The individual who looks out for genotypes other than his own will have a lowered likelihood of passing his genes on, and is thus selected against.

   Does this mean that an adaptation can evolve which, though benefiting the individual possessor, can threaten the whole population? It does. A predator population has a greater likelihood of avoiding extinction if each member strives to victimize only sick prey animals, those that could not survive anyway. In that way the prey population could be maintained at a larger size, ensuring a more secure food resource for the predators. Sadly, those individual predators who make no attempt to concentrate on sick prey by passing up healthy ones, and who grab whatever is convenient, will leave more genes behind, on the average, than their apparently more conscientious neighbors. Selection will favor the dastardly to the detriment of the population.

   Some definitions of group selection pose this kind of problem: Imagine two separate predator populations. In one you have selfish individuals devouring anything they can get hold of, whereas in the other population *all* individuals favor sick prey to healthy ones. Then suppose that the first population becomes extinct while the second persists. Here is group selection, perhaps. But is it selection *for* the group? If the surviving population had included alternatives in which some individuals focused on sick prey while others did not, it would have gone the same route as the extinct population. This, then, is less an example of selection that an example of what might happen when selection cannot act because of a lack of alternatives.

2. A nice proof of this is given in Lotka (1956).

3. It has always seemed odd to me that Fisher gave so little credit to Lotka; he is not even included in Fisher's "Works Cited." Fisher's book (1958, first published in 1930) is, nevertheless, a pioneering work that deserves all the tribute traditionally given it. It is brilliant but frustratingly recondite. Lotka's contribution (1956, first published in 1924) is somewhat easier reading but is less advanced in the area of population. Lotka, by the way, drew heavily from the work of previous thinkers, but he did not resist identifying them.

4. Anderson, Hassinger, and Dalrymple (1971) show that in a wild population of salamanders there is increased mortality at certain critical stages in individual ontogeny.

5. A related possibility deals with what is being called a "critical period." A newborn mammal or a newly metamorphosed insect may be facing a critical moment in life in that new systems are coming into play, taking on crucial responsibilities. The activities of any one system may inflict stresses on others. In time, each system may adjust to the workings of the other and, when this occurs, the critical period will have passed. On the other hand, it is possible that in some of these kinds of instances, no later adjustment can occur. In the purest sense of this latter possibility, birth or metamorphosis is more a testing period than a critical period that will shortly pass. As a test, some animals pass while others fail. In the extreme but unlikely case that there are simply those individuals who will almost

certainly pass and those who will fail with equal assurance, little or no selective advantage would accrue to concentrating adaptive strengths at such periods. It seems far more reasonable, however, to view these periods as times of *both* adjustment and trial. The ability of physiological systems to acquire harmonious adjustment among themselves has been documented in diverse instances, and the failure during difficult times of individuals with impaired capability of adjustment is more than a reasonable expectation.

6. See also Hamilton (1966), Guthrie (1969), and Emlen (1970).

7. One of the most puzzling examples of delayed reproduction is provided by the several species of periodical cicadas. These insects postpone reproduction for 12 or 16 years (depending upon species), to produce explosive emergences of reproducing adults every 13 or 17 years (Lloyd and Dybas, 1966a, 1966b).

8. More recently Gadgil and Bossert (1970) have examined the implications of the timing of reproductive efforts in the life history schedule. Their work presents a more detailed consideration than Cole's pioneering analysis, and notes some problems connected with ignoring the death rate in the calculation of r.

9. This "connection" between the reproductive-value curve and either the adaptive index or the age-specific mortality is incomplete. Hamilton (1966) points out that Fisher's statement is too broad and does not apply to ages beyond parental care. He suggests that an expression of the sort

$$w_x = \int_x^\infty e^{-rt} l_t b_t \, dt$$

is a more appropriate index of selection than is the reproductive value. This equation measures the expectation of offspring beyond age x and thus produces the same curve as our u_y in Figure 6.

However, during the ages of parental care, the state of survival of the offspring *is* an important factor since offspring lost sufficiently early can be replaced by the parents. Thus, during this period, the adaptive level should change according to a survival factor such as e^{rx}/l_x, which, when multiplied by w_x, gives Fisher's v_x/v_0. This would account for the high infant mortality seen in many species, and Hamilton notes that the early death rate could reflect an advantage to the parents in terms of replacement of lost offspring. The course of the adaptive index used herein roughly suggests the piecing together of two curves: the reproductive value up to the end of parental care, and Hamilton's w_x beyond it.

10. It hardly needs pointing out that the current chapter provides nothing like an appropriate background in population biology. There are several good books on the subject, including the clearly written, gently paced "primer" by Wilson and Bossert (1971). Their book is a good place to start, and it provides references and guides to more advanced material.

4

EVOLVING
THE IMPROBABLE

In the traditional, neo-Darwinian view, the concept that complex structures are evolved slowly, in small steps, is scarcely to be questioned. To Richard Goldschmidt (1938, 1940) and Otto Schindewolf (1950) it was indigestible. Schindewolf was convinced that the fossil record supported a view that new, major evolutionary steps occurred with jarring suddenness. Goldschmidt was stunned by the complexity of organic adaptations, the tight interactions, connections, coordinations, and symphonic cooperations among their components, and concluded that these complexities arose all at once, not little by little. He reasoned that the internal harmony of these adaptations was so complicated and crucial to the organism's viability that without all their integrative relationships present, they would not function at all. Thus he could not totally accept the concept of gradual, neo-Darwinian evolution, since the slow, stepwise method could not conserve the functional interdependence of each facet of a complex adaptation with all other facets. In Goldschmidt's eye, complexly interacting adaptations came into existence all at once or not at all.

Goldschmidt wrote extensively on this aspect of evolution. He proposed that major evolutionary changes took place by major mutations, "macromutations" or, as he often referred to them, "systemic mutations," which involved a major revision of the genome sufficient to alter an entire phenotypic system. The consequences of such a dramatic genetic upheaval were, on the phenotypic level, a new, hitherto unseen creature, a monster. According to Goldschmidt, some of these monsters would possess, by luck, the potential to survive. These he termed "hopeful monsters." If the hopeful monster actually did survive, and reproduced little monsters who themselves, when they came of reproductive age, could accomplish the same feat, a new type of creature would have evolved. From that point minor evolutionary changes could occur by traditional neo-Darwinian selection of small differences, and an adaptive radiation of many species would occur. At a future time another hopeful monster would appear, succeed biologically, and then an additional novel, basic type of creature would take its place in evolutionary history.

As one might readily predict, Goldschmidt's scientific colleagues eyed these notions as a geographer might regard the Flat Earth Society. It must have disturbed many, however, that these ideas were expounded by a person of such gigantic stature, for Goldschmidt was one of the century's leading experimental geneticists. No one stopped to consider whether in all of Goldschmidt's assailable propositions, there existed anything worth thinking about. There was no time for such consideration as long as there was so much merry mayhem to be carried out. In my university classes the name "Goldschmidt" was always introduced as a kind of biological "in" joke, and all we students laughed and snickered dutifully to prove that we were not guilty of either ignorance or heresy.

My present view is that in many things Goldschmidt was very wrong, but in others he was very right. On the negative side, I do not visualize animal evolution as having occurred through a series of phenotypic saltations. It is of concern to me that a wholesale reorganization of an individual's genome might well prevent a fruitful mating with a normal member of the population. The other possibility, that of two identical hopeful monsters appearing at the same time, finding each other, liking each other, and raising a family (if they happen to be oppositely sexed), stretches even gullible credulity to the snapping point. I am concerned that a brand new, untested biological form might not successfully compete with "normal" members of its population, who have evolved carefully and precisely for the ecological niche they occupy. And yet, even after admitting to all these difficulties, I can see that Goldschmidt has called attention to the very problem of evolving the improbable. He has emphasized the integrated nature of complex adaptations and hinted at the rules that must obtain in evolutionary transitions from one complex system to another. And he has produced what might be a most prescient essay on the way the improbable might evolve, by means that could hardly offend the most fastidious neo-Darwinian.

Among the more constructive criticisms of Goldschmidt's work is a paper by Mayr (1960), which, rather than being aimed at Goldschmidt's views directly, assigns for itself the task of determining how novel adaptations come into being. Mayr concludes that old structures may be modified in degrees to attain new functions. In that manner selection will be diverted to operate along new paths, which can lead to the formation of novel adaptations. The attainment of these adaptations is seen as gradual, arriving by the addition of tiny modifications until, eventually, a complexly integrated system emerges.

Certainly this is a reasonable view, and many—I dare say most—evolutionary biologists are at home with it. A

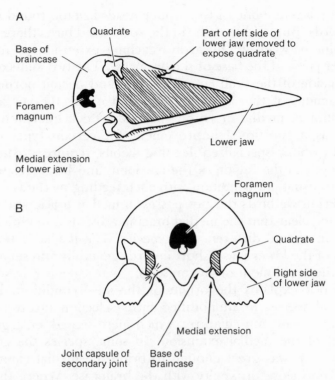

A. Ventral and B. posterior views of a bird skull possessing a **15** secondary jaw joint. (Modified from Bock, 1959.)

recent paper by Bock (1970) gives further comment on the origin of macroevolutionary events through microevolutionary changes. And in an earlier study (1959) he quite convincingly demonstrated the way that a new, adaptational shift can arise through gradual evolutionary change.

There Bock noted that, in certain species of birds, the lower jaw forms a double joint structure on each side of the head (Figure 15). There is, on each side, the typical joint between the (articular bone of the) lower jaw and the (quadrate bone of the) upper head skeleton found in all birds. But, additionally, in the specialized birds there is a joint between an inwardly reaching extension of the lower jaw and the base of the braincase. The two joints on each side of the skull lie side by side and permit normal movement of the jaw. The new joint offers the jaw an additional, medial brace on each side. Bock interpreted this as a functional improvement for certain types of birds whose specialized feeding habits require greater strength in the region of the jaw joint, and he correlates the unusual jaw condition with such feeding methods.

Primitive birds did not possess a medial brace, and it is quite clear that the medial bracing arose as a modification in several different bird species. The medial extension of the lower jaw is short in birds generally and serves for the attachment of jaw muscles.

Bock explains the origin of this new condition. In several species of unmodified birds selection favored a stronger jaw musculature, which necessitated enlargement of the medial extension. In some species the enlargement was great enough to bring the medial extension into close proximity with the braincase. Where this occurred, selection could now favor the formation of a secondary joint, a new structure with a new function guided by a type of selection pressure that did not—even could not—have existed before.

In some ways the secondary jaw joint is a major, evolutionary novelty; in some others it is not. Already present in the evolving birds was the capacity of bones and their associated musculature to adjust to one another. And the formation of the medial joint itself could have arisen quite easily, for, as Bock points out, bones and their surrounding tissues possess the ability to fashion a complexly structured movable joint (termed a pseudoarthrosis) where two bony elements are brought into moving contact. Considering these facts it could be said that the secondary joint, though an evolutionary novelty, did not require a very major change in the morphogenetic foundation of the organism. The major adaptive feature involved was the bone–muscle system, whose capacity for complex interaction permitted the evolution of the medial bracing device, but which had already evolved before birds appeared on earth [1].

Of course there is the possibility that had we the evidence to analyze the origin of the morphogenetic capacities of bone–muscle systems, we would find the acquisition of such capability a simple step from some immediately ancestral condition. Moreover, we might discover, in this case as in that preceding, that although the acquisition of this capacity did not involve a very major evolutionary alteration, it depended upon the existence of another, special morphogenetic ability already present in the ancestor.

The essence I wish to wring from all of this is that any major evolutionary step could have been achieved by gradual, rather than saltatory changes. But still it does not follow that all evolutionary changes are equally probable. It should be easier to evolve a modified color pattern on a butterfly's wing than a different wing shape. A change in color pattern could alter the butterfly's visibility to predators, its appeal to the opposite sex, and so on, while a change in shape could involve all those alterations plus a change in flight dynamics which, alone, might produce clumsy flying unless there were concomitant alterations in overall size, body proportions, flight musculature, and nervous control. Evolutionary changes that involve the acquisition or alteration of major adaptations are relatively improbable, and it is this fact which places certain constraints on the evolution of complex characters.

Perhaps nothing more at this time dare be said about complex adaptations. That might be quite enough, at least for a start. It might, in recommending greater attention to the restrictions connected to the improbability, get inquiries moving along the right path.

The improbability is traceable to two sources. There is the difficulty of the "right" mutations appearing in the population. Relatively few mutations are likely to improve a biological system, and when that system involves a complex interaction of many components, there should be an even greater unlikelihood of improvement. The second problem is the selection of the mutant once it appears.

If the first problem is examined against the background of the largest evolutionary changes that have actually occurred, it could seem remarkable that there was enough time in all the history of life to permit them. It is easy to get the impression that several specific sorts of mutations, affecting certain characters in certain ways, had to be combined within a single individual before favorable selection could take hold. This assumption is quite explicit in the evolutionary time puzzle with which I ended Chapter One. That puzzle was based on some very questionable premises, which should now be dissected.

In the first place I assumed that there was, in the fish–amphibian transition, but one way to produce a terrestrial tetrapod. There, in fact, must have existed several ways, some utilizing different fish ancestors (non-rhipidistians). The illogic of the game's assumption is a little like marking off a circle into 360 numbered segments, twirling a pointer which, when it stops at the (for example) 123-degree mark, elicits the comment that a most astonishing phenomenon has just taken place—the chance that the pointer would stop at 123 is no more than 1 in 360, a veritable miracle!

A second fault in the game is that it took into account but one population of rhipidistian crossopterygian fishes. Even if we insist that terrestrial vertebrates had to arise from rhipidistian fishes, there were undoubtedly many closely related rhipidistian species alive at the same time. This fact must certainly improve the probability of the initial stages in the evolutionary transition.

Third, the necessity that many characters be altered simultaneously may not be unreasonable, but these character modifications might not be independently controlled, as was assumed [2].

Finally, and as an alternative to the third point, it may not be necessary after all to combine a number of modified characters in the same individual. Perhaps one feature can change, which then opens the way for a change in another feature to become adaptive, and so forth. Possibly each added modification can integrate with those preceding, to gradually construct a highly complex, interacting system of functional components.

These four points seem to me to be the major objections to the evolutionary time puzzle, but the last two items demand closer scrutiny. It is important to notice that in point four, although modifications are added one at a time, the proportion of suitable modifications for each successive step will rapidly become small. As more characters must be integrated by each new change, it should be reasonable to expect that appropriate changes that will coordinate with larger numbers of components will be harder to come by. Once integrating response patterns have been achieved, the origin of additional components may become easier. But in the process of building the integration, the new components will have to be added with painful caution.

An associated problem, worsening the difficulty, is that if the genetic foundation of each modification begins as a rare gene in the population, as would occur in the case of a new mutation, the chances of accidental loss will often be exceedingly high. It has been demonstrated by Kimura and Ohta (1971) that even with positive selection value, the gene will most likely vanish from the population unless the selective value is quite high [3]. But if the selective value is high, this will often mean that the gene's effect on the phenotype is large. And if it *is* large, the probability that it will not have a harmful effect becomes minute indeed. Kimura and Ohta note that this set of facts places an upper limit on evolutionary rates. If the same mutation recurs in each generation, the mutant gene has a much improved (although usually small) chance of surviving, but the number of generations required for it to spread through the entire population could be enormous.

The third point challenging the evolutionary time puzzle is that the appropriate concurrence of changes that may be needed in a major evolutionary transition may not be independently controlled [4]. For instance, in the secondary jaw joint in birds it seems that the specialized mode of life of some species required changes in the size of the jaw muscles, in the length of the medial extension, and in the presence of a medial joint. Yet it is likely these are not independent matters. Experiments show that an increase in musculature can cause a corresponding growth in its bony attachments [5]; in birds, the increased muscle mass favored by selection was quite likely correlated with a longer medial extension, which, when sufficiently long, directly elicited production of a joint. The key factor involved is so very much the integrative capacity of the components that the actual presence or absence of the secondary joint itself reduces to an insignificant difference. The much more urgent question asks how this integrative ability evolved. It is thus important to distinguish between adaptive changes that only seem to be profound, and those which require profound alterations in the ways in which the components of a system adjust and influence one another.

It is the latter kind of evolutionary change that intrigued Goldschmidt, although he continually confused the former type with it [6]. He was certainly aware of the evolutionary potentials of integrated systems to form new adaptive modifications without requiring profound reorganization of the genome. But he could not leave alone the idea that most large phenotypic transitions in evolution stemmed from changes appearing with—quite literally—monstrous suddenness.

Recent authors who have become interested in the problems of major evolutionary steps are giving increasing attention to the major mutational events that must accompany significant evolutionary modification. The interesting thoughts formulated by such investigators as Rendel (1965), the Goins (1968), Stebbins (1968), Ohno (1970), and Britten and Davidson (1971) lay emphasis on the general topic of large genetic changes. Their theories differ in several details, but all suggest that a duplication of genetic material could provide a first step in major evolutionary shifts [7]. Ohno's book gives a pleasantly styled treatment of this idea.

The mechanism of this duplication could involve polyploidy, which doubles the entire genome to produce, in the case of diploids, a tetraploid. Or it could involve tandem duplication, the repetition of a certain DNA sequence as could occur, perhaps, from unequal crossing over. Among vertebrate animals, polyploidy is nearly impossible for all but some fishes and amphibians whose sex-determination mechanisms will not be disrupted by multiplication of the entire chromosome set (Ohno, 1970).

The gene-duplication concept suggests, a bit dimly at this time, how a major evolutionary transition could be effected. The argument runs something like this: A duplication produces two, functionally equal genes. If one of these continues to act in the formation of the original gene product, the other may mutate without depriving the organism of some vital function. A beneficial mutation that appears in this fashion may supersede the original gene function or may add a new gene-controlled activity to the old. In any case, the continuing activity of one of the duplicated genes in its original role is supposed to permit a smooth, undisrupted, adaptive transition, even though a drastically different system is being molded.

The problem of transition does not, unfortunately, really vanish completely. How, after all, will the duplicated gene, undergoing mutation, not compete with the unmodified gene? How will it fail to unbalance the genetic control as might any mutation? As long as the organism depends upon the original gene activity, how can the acquisition of a large alteration in the duplicated gene permit a smooth, adaptive transition?

It seems to me that what is wanted is a means of preventing the mutating, duplicated gene from expressing itself except occasionally, and then only slightly. This would not only allow the appearance of mutations without major disruption of the phenotype but would, through subdued expression of the mutation, permit selection to guide the mutation gently into increasingly adaptive genetic surroundings, by choosing for it the most harmonious recombinations or additional mutations. This action of selection would, of course, occur by the advantage of those mutant individuals with the most adaptively complementary genotype over mutants without such genetic backgrounds. But until there are sufficient numbers of mutant genes of this sort to be placed in several genetic backgrounds, and until their numbers are sufficient to escape random loss from the population, their fate could well be determined by genetic drift unless each mutant tended to have a very high fitness.

A system very much like this was proposed over a quarter century ago by Goldschmidt (1946). Goldschmidt's essay was intended to show how mutants having major phenotypic effects could produce only subtle changes and how, therefore, mutations with the potential for tremendous adaptive revisions could be spread through a population by favorable selection. In his terms, many new mutations possess a variable expressivity, depending upon the modifying effect of other genes in the genome. Hence a mutant may be completely silent, partially expressed, or fully manifested. I have but slightly modified Goldschmidt's graphical model for Figure 16.

Such a mechanism could allow many mutants to go undetected by selection. On occasion, some individuals who possess mutations will, owing to partial expressivity, exhibit a slightly modified phenotype. If the expressivity is small, and if one of a set of duplicated genes continues in its original function, the overall phenotypic effect could amount to no more than a subtle deviation from the "normal" type. Even if these deviant forms are less viable than the normal phenotypes, selection against them will be mild rather than vigorous. The mutants,

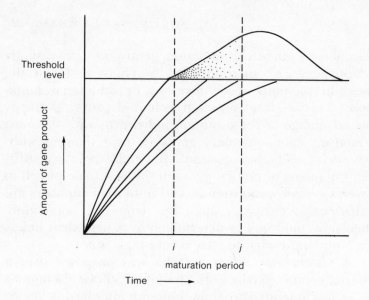

Threshold level

Amount of gene product

i *j*

maturation period

Time ⟶

16 Goldschmidt's model (1946). Within a population a number of individuals possess a mutation affecting a particular organ system. The system begins its final development toward completion at stage *i*, and achieves determined form (i.e., it reaches a point after which it is not significantly affected by a mutant gene product) at stage *j*. A certain quantity of gene product is necessary to reach a threshold level, at which point the mutant gene can be phenotypicaly expressed. If the rate of gene product formation is governed by other genes in the genome of each individual, differences in the rate could occur which result in variation in the time the threshold is reached. This, in turn, determines the degree of the mutation's expression in each individual. The bell-shaped normal curve in the diagram represents those individuals in a population who possess the mutation. The bell-shaped curve's abscissa is the stage at which the threshold amount of gene product is reached in the group. It is assumed for simplicity that the time required to attain threshold levels is normally distributed. Shaded regions of the curve indicate the degree of mutant expression in individuals that comprise the group.

rather than being completely exterminated, will be rated by selection as more less acceptable according to the relative effect of each in the context of its own genetic surroundings. Some of the surviving mutations will be completely masked in the next generation, while others will be moderately expressed.

This kind of mechanism could maintain a partially hidden reservoir of mutant genes, exposing them just often enough for selection to promote slow improvement of the genetic systems containing mutations. In Goldschmidt's view this could allow the introduction of a major adaptive innovation whose earliest phenotypic appearance—and interaction with selection—is little different from gene mutations of far less significance. Even though none of these suggestions may persist as future research provides more information, Goldschmidt is due credit for a macromutational approach that is quite neo-Darwinian in implication [8].

The gene-duplication theory has been involved in recent speculations by some that structural differences between certain well-studied biochemical systems (e.g., hemoglobin) are due to nonadaptive, random fixation of neutral mutations. Theoretical foundations of the random drift and fixation of neutral mutant genes are extensively set forth by Kimura and Ohta [9].

No matter what mechanisms are proposed to explain genetic alterations having major adaptive effects, there is a rough scale of probability along which the larger adaptive mutational changes will measure as the more unlikely. We shall revisit this vitally significant matter later in the chapter. Before doing that it is best we turn an eye toward the second problem in the evolution of major change, that of selection itself once appropriate mutants appear in the population.

There seemed once to have been not a great deal of concern whether, implicit in the results of laboratory selection experiments, any clues were revealed pertaining to an upper limit of evolutionary rates in natural populations.

The atmosphere changed quite decisively in 1957 when J. B. S. Haldane published his now famous paper, "The cost of natural selection." Since then numerous articles on the subject have appeared, and the question of upper limits on evolutionary rates has, with varying degrees of hospitality, been welcomed into the discussions.

Haldane's approach is quite intriguing. He notes that for selection of survival characters to occur there must be differential mortality, and he asks how much "cost" in terms of mortality is necessary to exchange one allele for another in a population. His model is applicable to the situation where a rare gene suddenly becomes selectively favored due to a change in the environment of the species, or where a new mutation, having selective favor, appears in the population at low frequency. Arising from Haldane's calculations are two interesting results: When the selective advantage is low (less than about 10 percent), the cost is hardly dependent at all on the magnitude of the advantage. And the cost per gene substitution tends to be very high.

Implicit in the results is that it often takes many times the number of population members in actual deaths before a gene substitution can be completed. Haldane estimated that, in an average case, if the population size were to remain nearly constant over the period of gene substitution, about 30 times the number of population members must die. In a teleological sense this leaves the population with an unhappy choice: Does it evolve fairly rapidly and risk a drastic reduction in size and the flirtation with extinction due to random processes, or does it maintain a large size but evolve very, very slowly? These alternatives are popularly framed in the term "Haldane's dilemma." Its most striking implication is that there are grave difficulties in selecting for more than one gene at a time when each substitution extracts such a Draconian fee.

The cost concept and its supposed consequences have been both rejected [9] and accepted [10]. The controversy still glows and there has been some revision of terminology in recent years. Rather than "cost," the somewhat more realistic, although ponderous, "reproductive excess necessary to prevent extinction," or similar-sounding terms, are now employed. The revised terminology has the advantage of acknowledging that living things produce more progeny than can possibly survive, and that almost any population will always bear a burden of mortality whether gene substitution is occurring or not. This will be especially true when the population is at the saturation level [12]. Hence, during gene exchange due to selection, the *number* of dying population members may be unaffected, the selection taking place in terms of *which* ones will be eliminated [13].

Haldane's critics argue essentially that the conception is meaningless in terms of natural populations, and that genes can be substituted simultaneously at many loci. A basic ingredient of the arguments is a pivotal assumption regarding the way that selection works. The reasoning of King (1966), Milkman (1967), Maynard Smith (1968a), and Sved (1968) in particular emphasizes that since selection acts primarily on the individual, one must deal with the relative survival capabilities of individuals, not of particular genes. In any generation one can arrange all the individuals of a population along a scale of fitness,

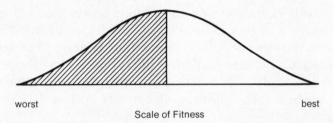

worst   best

Scale of Fitness

17 Frequency distribution of individual fitness in a population, illustrating the Maynard Smith–Sved truncation model. The shaded portion shows the individuals selectively removed in such a model, where only 50 percent of the population survives to reproductive age.

with those whose survival chances are the worst at one end, and at the other end those whose chances are best. In Figure 17 I have such a scale on the abscissa and assume that the entire population is normally distributed with respect to fitness. If the fitness of each individual is determined by his beneficial genes, and if we imagine that at many loci there are good and bad alternative alleles, it should follow that individuals toward the right end of the abscissa contain more favored genes than those on the left. Suppose now that the conditions under which the population exists are such that only half the members survive to reproductive age. If selection were perfectly efficient, it would remove only those individuals in the shaded half of the diagram, thus immediately boosting the frequencies of many genes present in individuals represented in the unshaded portion. Maynard Smith calculates that under the operation of truncation selection, where all the less-fit members up to a certain level of fitness are removed, the number of possible gene substitutions for a given rate of gene-frequency change can be a good many times greater than Haldane's model allows.

A major conceptual difference separates the outlook of Maynard Smith and Sved from that of Haldane and his supporters. In the Haldane view, most genes are independent and control different adaptive features: A single truncation diagram, as given in Figure 17, will not do, because it can refer to but one gene-controlled trait. If the traits are really independent, one would need a separate diagram for each different aspect of selection.

For instance, suppose that in a species of moth it became advantageous for it to alter its flight characteristics. Further pretend that there are several alleles at different loci, initially rare, and that each has a single effect as, say, improvement of wing shape, increase in stamina, strengthening of the flight muscles, and so forth. All these features are related to flying ability, and the moth with the greatest number of these genes will be the best flyer. Hence, when it comes to selection for flight performance, a truncation scheme will permit the simultaneous substitution of many genes. But now suppose there is also an advantage in acquiring a camouflage as protection from predators, and another in altering the mouthparts to take advantage of a new food resource. If the gene for mouthparts does not protect the moth from predation, and if the gene for camouflage does nothing for its flying ability, and if, in fact, these adaptive aspects are truly as independent of one another as they seem, a special truncation diagram would be needed for mouthparts, another for flying, and still another for camouflage.

This gives rise to a serious limitation on selection. While those moths with the worst mouthparts are being weeded out of the population, they take with them some favorable genes for flying and for camouflage. There is, according to this kind of argument, no selective basis by which they can be credited with the flying and camouflage genes, because these factors are independent of selection for improved mouthparts.

In the Haldane formulation the probabilities of survival after selection for each trait are, in effect, multiplied together so that the prospect of successfully enduring a number of independent selections diminishes enough to retard the spread of several genes at once. Maynard Smith, along with Sved, clearly pointed out the difference between the two modes of selection, suggesting that few traits are likely to be independent. Nei (1971), on the other hand, argues for the view of adaptive independence. I devoutly hope I shall not be expected to resolve this conflict here. I can only uncourageously agree with both sides that some traits are certainly independent, and some others are certainly not. For me, the great bulk of the possibilities seem to require careful attention before they will be properly allocated.

It is plausible that some kinds of new adaptive advantages governed by initially rare genes depend upon the presence of *several* specific genes in the same individual. In this imagined case the genes' effect upon one another is synergistic. Individual genes, or groups of genes smaller than the full complement, provide little or no fitness advantage. But very few individuals would be so lucky as to possess the entire complement, and much of the population's supply of these genes would be distributed in individuals ranged along the entire fitness scale. The increase in frequency of these genes by selection could thus be exceedingly slow. Although synergistic activities among genes are known [14], we lack evidence regarding the importance of such gene systems in evolution.

The controversies following Haldane's paper have been of particular value from our standpoint. They have so frequently touched on the problem of complex adaptations as to mold it into some vaguely suggestive shapes. For one thing, the Haldane model establishes that interdependent genes (except in some cases of synergism) are more easily selected than independent factors. One might predict that most favored genes will possess a wide range of adaptive effects, and that relatively few will affect but a single aspect of an animal. The prediction is perhaps secure in fact [15], although the fact itself could have other interpretations. If the interdependence is based only on the kind of synergistic relationship hypothesized alone, selection would be quite inefficient (but once such a system evolved it might be difficult for selection to dismantle at a future time).

There are other implications arising from the "cost" controversy. One approach to them is through consideration of a survivorship curve, such as was discussed in chapter two. The shape of the curve shown in Figure 18 is arbitrary, but about any reasonable pattern of survivorship will do for the arguments that follow.

I am supposing that each of several rather low frequency genes influences a survival trait A, which has no selective significance for the possessor until he reaches age a_1. At that age, and until age a_2, the genes exert a positive influence on survival and are thus selected for; no other trait is being selected during this period. At the age of selection there are about 80 percent of the original cohort still surviving, and by age a_2 only half of those remain. We might assume for realism that the entire population is at its saturation level and is neither increasing nor decreasing, a condition that is perhaps average in most natural cases.

If the effect of selection can be thought of as a truncation of the fitness-distribution curve of the cohort, then the most unfit 50 percent of the members entering

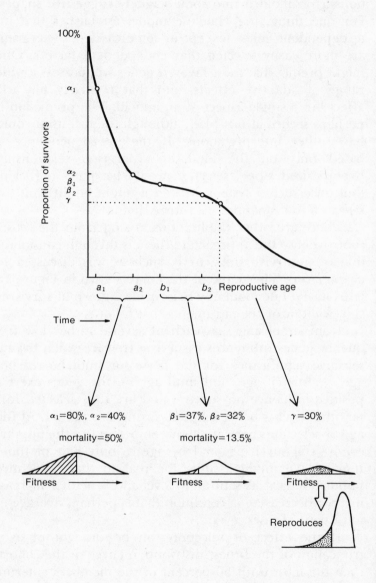

age a_1 will have been taken by natural selection by age a_2. Now in similar fashion another survival trait, B, independent of A, becomes significant from age b_1 to b_2, and is also controlled by genes at low frequency. Because of the shape of the survivorship curve, only 13.5 percent of the individuals are lost through mortality in the b_1–b_2 age interval. In this case the excess of individuals available for differential mortality is less than in the former, with the consequence that selection is potentially stronger in the case of the A trait. An appropriate alteration in the shape of the curve could have, of course, reversed matters, placing B in the position of greater potential selection.

If traits A and B, once they become selectively important, *remain* so throughout life, trait A can always be more strongly selected. Survivorship curves continually lose altitude and the earlier a trait acquires selective importance, the greater is the excess of individuals, as measured by mortality, available to selection [14].

The characters with the best chance of spreading through a population should often be those whose selective advantage is of longest duration. This is certainly obvious from other sorts of considerations. It is equally obvious that traits whose adaptive effect becomes manifested early in life will be favored, not only because of the increased span of selective activity that this implies,

Generalized model showing relationship of survivorship curve to upper limits of selective intensity at different ages. Mortality effects on individuals during the reproductive age are ignored for simplicity. The lower right curve shows the relative numbers of progeny corresponding to any point in the parental fitness scale (showing reproductive capability in this case) lying just above it, and is intended to show how many progeny the totality of parents of each given fitness value can produce.

but because of the high mortality in early life characteristic of many animal species.

Selection for reproductive success does not require mortality. Individuals with superior genes for fertility could leave several times more viable progeny than could others. Figure 18 shows the relatively few survivors of the original cohort reproducing their share of a new generation. I assumed that fitness for reproductive ability is continuous from good to bad, but in the figure I have stippled a portion of the parent's curve to suggest those individuals (in a hypothetical case) who will produce no more than one offspring apiece, and have left unstippled the region that represents parents who will do at least slightly better. The progeny curve is intended to show how very poor might be the relative contribution of the "stippled parents" to the gene pool.

If one or more genes related to a survival trait are *very* rare, they may be accidentally lost from the population despite a slight selective favor (Kimura and Ohta, 1971). In the case of a mutation that occurs only rarely, a population may go through several generations before a single mutant gene appears. Once it does, unless selection preserves the mutant very quickly, the population will probably lose it. But if the mutation is to have much hope of success, it should not only be favorable but strongly so. Yet, mutations that effect the phenotype in minor ways, if they have advantages at all, are apt to have only small, selective advantages. Mutations having larger phenotypic effects may be strongly advantageous or strongly disadvantageous. Large mutations with beneficial effects are extremely rare (Kimura and Ohta, 1971). What could be a vicious circle might be broken into by the very fact that although mutants of substantial magnitude and beneficial effect are terribly infrequent, they can enjoy strong, positive selection. Much more information on these matters must come into existence before speculations in this area cease to be dangerous.

The continual decimation of an original cohort reduces the proportion of older individuals in a population. Unless earlier favorable selection has preserved the rare mutants in a population, they might be hard to find in older cohorts. The significance of early selection must be tremendous in species that produce myriads of offspring which, by the reproductive age, are reduced to a mere handful. In early life the size of the cohort may be so enormous that even quite unlikely mutations may exist in at least a few members. With strong selection and luck, a very unusual mutant could be preserved [17].

Thus there is heavy significance attached to the time in life when a mutant will have an adaptive effect. Mutations with early beneficial effects have a far better chance of fixation in the population than those remaining selectively neutral until sometime later. This will be more true of advantageous mutations that have larger effects and which are, consequently, quite rare [18].

The implications of all these speculations, although by now obvious, are still curious. Adaptations for survival that affect younger animals should be easier to evolve than others. Evolutionary innovations for survival which are valued primarily by older animals (i.e., those closer to the onset of reproductive age) should be less frequent. Selection at older age levels will be no more than a veto on what is left from selection acting at an early age. Older individuals will have fewer selective choices than younger ones. Whatever might be of selective advantage in older ages would have had to pass an early selective screening. Thus evolutionary innovations that enhance survival will tend strongly to be those which are useful early in life;

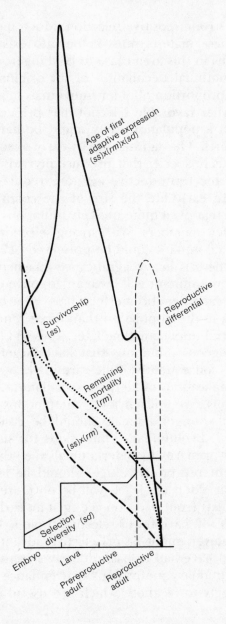

the adaptive themes in the adult have, in large and frequent measure, been prescribed in the young.

I have made several very arbitrary assumptions in drawing Figure 19. That fact has not ruined my hope that the graph provides substantial hints about the interaction of several factors in the evolution of a major adaptation. The survivorship curve is of arbitrary shape. It closely imitates the general shape for some species, though not others, but it is generally representative in that survivorship always decreases with time [19]. There is often a high early mortality in nature, and any cohort of individuals who are all born during the same period may lose much of its membership rather rapidly. A beneficial mutation of more than trivial effect will be exceedingly rare. Its chance of existence will depend directly upon the number of individuals present. The survivorship curve is thus a rough, relative measure of the chance that a rare mutant will be found among animals who have attained a certain age.

The curve for "remaining mortality" has been calculated from the survivorship curve as $(l_x - l_f)/l_x$, with l_x as the cohort's numbers at any given time and l_f the number left at the end of the reproductive period (the calculated curves in the figure are multiplied by constants so as to get them all into the available space). For survival characters, where population size is unaltered by the action of selection, the curve represents the upper level of selective intensity that individuals of a given age can experience.

The chances for an unusual, adaptive mutation to be secured by a population within a time unit will depend both upon its probability of appearance in the first place,

A representation of age-related factors bearing on the evolution of complex adaptations.

and its progress toward selective fixation in the second. I have very crudely hinted at the interaction of these two factors simply by multiplying them together to derive the curve $(ss) \times (rm)$.

To my mind, however, matters seem more complicated than I have shown so far. The organism is not exposed to the same degree of survival selection throughout its life. While fully admitting that an animal passing time in the egg or in the womb must be adapted to that embryonic mode of existence, I must acknowledge that it receives a degree of protection while doing so. The diversity of selective influences during this period is perhaps relatively small, and in a number of respects, the embryo can be buffered against many kinds of selective pressures. Through this time, the embryo's survival depends largely upon the special adaptations of its embryonic chamber, whether egg or mother, while its own structure enjoys considerable isolation from selective severity. Once released from its protective chamber, however, the individual finds himself in harsh exposure to the perils of nature.

The latter factor is represented by the curve labeled "selective diversity." I prefer to leave the concept intuitive and imperfectly defined, because it seems that any attempt to do more will add nothing we need for speculation at this level [20]. The assumptions used in drawing the curve are simple. The embryonic value is small (but increases slightly as the individual grows and encroaches upon the available space in its restricted quarters), then jumps to a high value at natality, which remains unchanged until the reproductive period, when it rises to a new level. A great deal of postnatal, parental care could, of course, impede the early rise of the curve.

When the "selective diversity" curve, the survivorship

curve, and the "remaining mortality" curve are multiplied together, the result is the curve $(ss) \times (rm) \times (sd)$. Theoretically, for a complex character, the curve shows the most likely age at which selection originally became important. The curve is thus intended to estimate the "age of first adaptive expression" of a character and is so designated in the figure. Many changes in our assumptions will not budge the high point of such a curve from the beginning of postnatal life. The steeper the survivorship curve, and the rarer the mutational event, the more significant this curve becomes. The tendency for the peak of the curve to correspond to early life is of no small importance. There are implications here for understanding the nature of complex adaptations. And if our general conclusions are, in the main, valid, they provide a clue as to how we might envision the sequence of changes that have occurred in particular phyletic lines of animals. The tendency for the peak of the "age of first adaptive expression" curve to occur early in life, when complex survival adaptations are involved, I shall designate as the "force of thematic prescription," for, as we have seen, the adaptive survival themes in animals tend to be prescribed by selection during early age. The most obvious place to seek evidence for thematic prescription of survival adaptations is in the numerous cases of species having complex life histories. There, in a single ontogeny, individuals possess two or more very different sets of survival adaptations in the free-living period of their lives. Are the most revolutionary adaptations to be found in adults? Or in the larvae? In a very great many cases the evidence points to the larvae. These cases include species whose patterns of metamorphosis list among those which are of the best known and most spectacular to be seen in the animal kingdom.

Tadpoles are very different animals from adult frogs. A tadpole's internal structure, mode of life, external form, diet, and so forth provide little hint of the kind of creature it will become. Neither are tadpoles very suggestive of other sorts of amphibian larvae. The adult frogs, however, although wonderfully specialized jumpers, resemble in important ways other amphibians, both living and fossil. It looks very much as though the great, major adaptive innovations in frog evolution involved the larvae more than the adult (see DeJongh, 1968).

Insect metamorphosis involves alterations of the most striking sort. Few dramas compare with the emergence of a brilliant butterfly from a cocoon which, only a few weeks earlier, enclosed a worm-like caterpillar. There are many life-history patterns in insects, including one in which no metamorphosis occurs, the newly hatched individual closely resembling the adult. This type of life history, termed "hemimetaboly," is usually regarded as primitive. The extreme metamorphosis seen in butterflies and many other insects, "holometaboly," is thought to be a more recent evolutionary innovation and may have been derived from a hemimetabolous life style.

Newly hatched holometabolous individuals are not in the least reminiscent of the adult form. They are, in fact, much more comparable to hemimetabolous embryos, and one theory has it that in the evolution of holometabolous insects, selection favored an earlier hatching time for the embryos. These "embryonic" larvae evolved various survival features, many of which are basic innovations. The pupal stage present in holometabolous forms may represent the last stages of primitive embryogenesis merged with what had primitively been postnatal preadult stages [21].

If these speculations on the evolution of insect metamorphosis are accurate, the earlier postnatal life stages, not the later ones, display the greater degree of adaptive novelty. In broad implications, the example seems to possess some generality. Sir Gavin De Beer's book (1958) provides pages of examples, drawn from many diverse animal groups, showing that major adaptive innovations quite usually make their first appearance in early life [22]. Our own phylum, the Chordata, quite probably arose by the evolution of new adaptations in a prechordate larva, which then were gradually retained, by action of further selection, into the adult stage [23].

Some adaptations are not, however, primarily for individual survival but for reproductive success and must be viewed a bit differently from those we have just considered. In Figure 19 I show a curve labeled "reproductive differential" arising at the beginning of the reproductive period and falling at its termination. The height and shape of such a curve evidently depend upon the species being regarded. Here, and only for convenience, I have drawn the peak of the curve to the same height as the initial point of the survivorship curve. It represents the potential maximum difference in viable offspring that one surviving individual can produce versus another. It is thus a crude estimate of the upper limit of selective advantage due to reproductive success. It is only coarsely analogous to the "remaining mortality" curve, which measures upper selective limits for survival features in terms of a ratio (of those to die versus present survivors); the "reproductive differential" is an arithmetic difference between some number of offspring and zero, the latter being the worst a would-be parent can do. The height of the reproductive differential curve can be very great because of the great intensity of selection on reproductive adaptations, although survivorship to reproductive age may be quite reduced.

Despite the potentially high selection for reproductive

characters, the availability of mutants at reproductive age has dwindled significantly. This should suggest that there will often be fewer kinds of major innovations in reproductive adaptations than in survival adaptations. I am not sure that this is really a fact of nature. The ornate plumage of birds of paradise, the complex mating displays of many insects, the antlers of deer, the remarkable nest-building abilities of weaver birds are some among many examples hinting at a notable diversity in reproductive characters. Yet, whether many of these represent adaptations of a basically novel sort, originally evolved for reproductive success per se, may be an unresolvable issue at the moment. The ornaments of birds of paradise are, after all, feathers that certainly evolved as thermal insulation and/or as an airfoil in some *Archaeopteryx*-like creature. And the complex territorial behavior of breeding birds may not represent a great evolutionary step once the neural machinery for other behaviors has evolved.

We should expect that expression of reproductive features, or any feature that is adaptively useless until later life, will be reduced or absent earlier in the postnatal life span. A structure whose benefits have not yet arrived is probably disadvantageous in the meantime.

Thematic prescription, like other hypotheses suggesting what will most probably happen under the given assumptions, is imperfect in that its general predictions cannot reasonably be expected to hold for every particular evolutionary case. Exceptional examples will undoubtedly exist, but as long as they amount to no more than an acceptable minority of cases, they need not trouble the broad implications of the concept. For instance, it is difficult to find congruence between the notion of thematic prescription and such matters as the ontogeny of wings in insects. In both hemimetabolous (where the newly hatched individuals resemble the adult) and holometabolous (newly hatched individuals attain adult form by a profound metamorphosis) insects the wings are not functional during the earliest, free-living stages. Until the wing is in working condition, it is represented as a wing pad or developing rudiment and, at least in general, bestows no selective favor upon the young possessor. At that stage it is quite likely disadvantageous to survival. The thematic prescription theory would predict that such major structures as insect wings were originally of selective advantage right after eclosion or hatching. But presently there is no evidence at all that this was the case. They perhaps have always been adaptations for later life but even at that, the selection pressure favoring their initial establishment in some ancient population could have been enormous. Wings, among other usages, enable the female to implement a choice of better egg-laying sites, and wings must have been subject to the tremendous selection available to reproductive adaptations profiled in Figure 19 [24].

When new, major features first show themselves in a population, their future will depend upon their selective advantage. That advantage will reflect how "perfect" the new machinery is and has to be. The perfection of adaptations is a difficult but pertinent subject, and is the topic of the next chapter.

NOTES

1. I have more to say on this topic in Chapters 7 and 8.
2. Again, there will be more on this in Chapters 7 and 8.
3. Kimura and Ohta (1971, p. 10) estimate that for every rare

mutant gene with a selective advantage of 0.5 percent which becomes fixed in the population, 99 are lost.

4. Simpson (1953), in his thoroughly stimulating and important book, provides an evolutionary time "game" based on tooth characters in horses. He concludes (pp. 109–110) that even with overall mutation rates as low as 10^{-6}, more than sufficient time was available for the appearance of many mutations in the evolution of the teeth of modern horses from the Eocene "dawn horse," *Hyracotherium.* The mutations are assumed by Simpson to be quite infrequent, but then some of the tooth characters in question are not major. Even so, the factors of random loss and possible long fixation time of beneficial mutations could greatly slow the evolutionary rates. But on the other hand, it might be that whatever tooth characters are involved are not all controlled by independent genetic units.

5. This is discussed at greater length in Chapters 7 and 8.

6. I have never been sure whether Goldschmidt discerned the difference between these kinds of changes or not. It seems to me that for each of his statements recognizing the difference, there lay a contradictory phrase within no greater distance than a paragraph or two. This problem in communication did not help to clarify Goldschmidt's argument for his readers. He himself said (Goldschmidt, 1955, p. 489), "I have been reproached for not having made it clear in my book *The Material Basis of Evolution* whether I was speaking of systemic mutation (scrambling of the chromosomal pattern) or of ordinary mutations of a macroevolutionary type, and of [sic] being confused myself on what I meant."

7. But see McLachlan (1972) for a critical view.

8. Goldschmidt did not himself favor a concept of gene duplication, although he considered it. The model he presented in 1946 was not based on duplication, and in 1955 (pp. 485–486) he said: "Bridges introduced the idea

that a duplicated gene ... may in time transform into a completely new gene. The presence of the old one permits such a development of a new one without harm to the organism. I have always felt that this idea is very crude and, in addition, contrary to all we know about the action of the gene, if we argue now within the classic theory of the gene." And a bit later: "In all cases known, the assumed duplicate has generally the same action, and is different only to the same extent as another allele."

9. Kimura and Ohta (1971); but there are critics of the view, for example, Uzzell and Corbin (1972).

10. For example, Brues (1964), Moran (1970), Maynard Smith (1968a), Sved (1968), Van Valen (1963).

11. Crow (1970), Felsenstein (1971), Hecht (1965), Kimura and Ohta (1971), Nei (1971), O'Donald (1969), Van Valen (1965a).

12. See Note 5 of Chapter 2.

13. Several calculation methods have been suggested which lead to results similar to Haldane's. For an illustration of the way the idea can be developed I present here an approach based on that given by Crow (1970) and by Crow and Kimura (1970). For simplicity's sake I shall deal with a haploid species. The diploid case is mathematically tedious and does not add much to a general appreciation of the concept's mechanics.

Suppose that in a haploid population there are two alleles, A and a, at a particular locus. Their initial frequencies are respectively, p_0 and q_0, so that, since these are the only alleles of that locus present in the population, they sum to $p_0 + q_0 = 1$. Suppose further that A begins with a low-frequency p_0, and, for whatever reason, now enjoys a selective favor over the alternate, a. Let us say that the disadvantage of allele a can be measured by a fraction s (the selection coefficient) such that the probability of survival by an a individual is the amount s less than the survival prospects of an A type. Hence the fitness of A individuals relative to a's is the ratio $1:1 - s$. This is summarized as follows:

(1) Haploid genotypes: A a

(2) Initial frequencies: p_0 $q_0 = 1 - p_0$

(3) Relative fitnesses: 1 $1 - s$

(4) Relative contributions to the next generation, (row 2) × (row 3): $p_0 \times 1 = p_0$ $q_0 \times (1 - s)$.

The new frequency of A in the next generation will thus be

$$p_1 = \frac{p_0}{p_0 + q_0(1 - s)} = \frac{p_0}{1 - sq_0} ,$$

the expression on the right following after substitution of $1 - q_0$ for p_0. The change in gene frequency in a single generation is the difference between p_1 and p_0, which is

$$\Delta p = \frac{p_0}{1 - sq_0} - p_0 = \frac{p_0 s q_0}{1 - sq_0} .$$

When the level of selection is low (s is small) the term $1 - sq_0$ will not be much different from unity. The change in p can thus be estimated as

$$\Delta p = p_0 q_0 s . \tag{1}$$

Let us leave this for just a moment and note that in any generation t, the proportion of deaths required for selection is simply sq_t. This can be used as a rough estimate of the *rate* of selective mortality and set equal to dC/dt, letting C be the "cost." The total cost is obtained by summing over all generations during the gene substitution to obtain the integral

$$C = \int_0^\infty sq_t \, dt . \tag{2}$$

Values given by Crow and Kimura (1970, p. 246) for an algebraic summation approximating (2) above (and which is derived in Crow, 1970) are given below. The initial gene frequency is set constant at 0.01.

s	1.00	0.99	0.50	0.10	0.01	limit
C	99	52	6.2	4.8	4.63	4.61

The bracketed region shows that at $s \leqslant 0.10$, the cost varies hardly at all with s. In this region the cost or number of deaths required is nearly constant at approximately 4.6 to 4.8 times the average population size over the time considered. Crow (1970), Crow and Kimura (1970), Nei (1971), and Felsenstein (1971) all prefer to measure the cost not in deaths but as a reproductive excess required to keep the population from going toward extinction. When the excess is insufficient, rather than the population diminishing below saturation level, it is much more likely that the rate of gene substitution will be lowered (Kimura and Crow, 1969).

The cost can be seen to depend entirely upon the initial gene frequency p_0 as long as s is small. To demonstrate this, we observe that equation (1) is roughly equivalent to the rate of gene frequency change in the continuous case dp/dt, and thus the approximation

$$\Delta p = p_t q_t s \approx \frac{dp}{dt} \tag{3}$$

can be written and substituted in equation (2), whereby we now have the solution,

$$C = \int_{p_0}^{1} \frac{sq}{psq} \, dp = \int_{p_0}^{1} \frac{dp}{p} = -\log_e p_0.$$

In this haploid example the cost, as we have already seen, ranges up to about 4.6 when $p_0 = 0.01$. In diploids the cost can be very much higher and increases as the initial gene frequency is lower, but also increases with the degree of dominance of the disfavored allele. Haldane (1957) chose 30 as a rough, average cost per gene substitution in diploid species.

14. See Mukai (1969) and also Crow (1970) and Nei (1971) for short discussions.

15. The impressive integrations and interweavings of components in epigenesis is sufficient evidence in itself that very many genes code for products that control the formation of still other components in development. Hence the adaptive significance of each such gene is remarkably broad.

16. The essence of the model proposed here is that, in most populations the mortality will tend to be unchanged by selection. Thus the model suggests an upper limit to the magnitude of selection, s, based on mortality. An alternative view would hold that the mortality, in part is *due* to selection. This would be the case in traits not related to adaptations for competitive selection (e.g., a sufficiently large decrease in rainfall could reduce, or even extinguish, certain populations). In such cases the selection coefficient could nearly control mortality directly. The proposed model applies best to the first situation.

17. The chance that a single mutant will ultimately spread through a population, rather than be lost, is roughly proportional to twice its selective advantage ($2s$) (see Kimura and Ohta, 1971, p. 10). Where the selective advantage is very low, $|2Ns| << 1$, with Ns being the product of the effective number of breeding individuals in the population and the selective coefficient; the probability of fixation of a mutant gene is given by Kimura and Maruyama (1969) as $p + Nsp(1 - p)$, where p is the initial frequency of the mutant gene. Also see Ewens (1970) for a discussion of this formula.

18. This model, based on the availability of rare mutants for selection, leads to the same conclusion as the immediately previous model, which examined the effectiveness of selection in terms of excess mortality.

19. See Deevey (1947).

20. Various measures of diversity have been applied to a quite different biological problem, that of analyzing species diversity in ecological communities. An often-favored measure is based on the Shannon and Weaver information content formula,

$$H = - \sum_{i=1}^{n} p_i(\log p_i).$$

If we needed to adopt a formulation for selective diversity, we might experiment with H defining p_i as measuring the intensity and direction of selection on the ith character.

21. Berlese (1913) is credited with the evolutionary interpretation of holometabolous larvae. Hinton (1948) considered the pupal instar comparable to an adult stage of hemimetabolous insects. A discussion of the two theories is given by Fox and Fox (1964).

22. See Istock (1967) for a theoretical consideration of selective factors maintaining complex life cycles.

23. The argument for thematic prescription may seem at first in conflict with the speculations (developed in Chapter 3) that in ontogeny, adaptive strengths will tend to be delayed until toward the end of parental care. The apparent inconsistency is not real. There will be a strong tendency for major innovations to be evolved for early stages of postnatal life. If there is postnatal parental care, whatever weaknesses the innovations possess will be scheduled— when selectively feasible—for earliest expression. The operational basis of an adaptation is not to kill the animal at an early stage, but to preserve it; but if the new, "preserving" adaptation is imperfect, then it is best to have the worst of the imperfection done with as soon as possible during the period of parental care.

24. This does not mean that insect wings could not have evolved initially as survival structures useful in very young individuals. According to current views the earliest insects

were wingless. The first wings could have evolved from body wall ridges (paranota) useful in gliding. Although we lack definitive evidence suggesting that these ridges, and their later modifications, were important in young individuals, there is no good reason to dismiss the possibility. Once wings have evolved, and all members of an insect population have them, the door is then open for future evolution where wing function can become suppressed in the young. For if the wings have a very important reproductive role in adults, granting winged parents a tremendous reproductive advantage, the adaptation can be preserved. Prolonging developmental time of wings, once wings have appeared in evolution, can better adapt them for the reproductive-related role of egg dispersal. Wing rudiments without immediate selective advantage in young insects might have a negative survival effect in early life. But as long as the disadvantage is not so enormous as to permit the survival of *only* the few genetic novelties with missing or reduced wing rudiments, those individuals with well-developed wings, and who escaped selection's screening to survive to parenthood, will contribute a huge share of "winged genes" to the next generation. It seems to me possible to test this hypothesis through study of wild insect populations. For a general account of wing evolution in insects, see, for example, Fox and Fox (1964) and Alexander and Brown (1963) for contrasting thoughts.

5

THE "PERFECTION" OF ORGANIC MACHINERY

From Babylonian times through many years thereafter the hour of the day was reckoned from the sun by a precise instrument, the sundial. The rules of sundial manufacture became quickly appreciated and practiced, and any deviation from them resulted in an inferior instrument. On overcast days it did not matter how recklessly the simple but precise mathematical rules of sundial construction had been violated. And it certainly did not matter at night.

There were, however, primitive but ingenious "water" clocks—run by a water or sand gravity flow—during most of that early period. These contrivances, although having something to say after sundown, lacked a bit of versatility in where they could be located, and were not as accurate as sundials. The first mechanical clock did not appear until roughly 1360 [1]. Known as the De Vick clock, it possessed definite advantages over earlier timepieces.

The De Vick clock ran by a weight drive, with a verge escapement and crown wheel, and thus it did not depend on the presence of water or sand reservoirs (it was even more accurate than "water" clocks) and did not depend upon sunlight. In a sense, it was the prototype of clocks to follow, and the advantages of this apparatus were gratefully enough accepted that, although other mechanical clocks appeared, their design remained little different from the De Vick clock for nearly 300 years.

But the De Vick clock and its imitators were noteworthy in some negative ways. The De Vick clock (which had one hand, not two as in modern clocks) was a poor time keeper compared with a good sundial. In fact, it did not keep time much nearer than two hours a day. By modern standards such performance is terrible. It suggests, although real information is lacking, that certain parts of this clock could have been slightly (but not necessarily carefully) altered in size and shape without having made matters much worse. Thus a precise and accurate device, the sundial, gave way to the imprecise mechanical clock in the evolutionary history of timepieces.

In regarding animals, it is easy to wonder how their adaptations should be viewed. Are they as perfect as a sundial, or as imprecise as the De Vick clock? Such a question could be a little hollow unless one can better define what ought be meant by "precision" or "perfection" in some convenient biological sense. Even without these necessary criteria, it is easy to get the feeling that animal adaptations perform so beautifully that each animal species is about as perfect as it can get (whatever that might mean—we can easily note such obvious imperfections as the requirement of frequent intervals of sleep). That some birds can fly at 100 miles per hour, that dolphins can swim at the respectable speed of 20 knots, that bone is stronger for its weight than many common building materials, and that examples such as these can be recited to the brink of boredom are all impressive facts. They are impressive because they relate to feats that man himself has envied and admired in the progress of technology.

To make comparison between engineered products and organic adaptations is a sweet temptation that many biologists have not struggled to resist. In certain fields, biomechanics for example, living systems are examined in terms of engineering techniques in an attempt to understand how the systems work and why they are built as they are. This is not an easy task, and certain of the difficulties point to broader problems in comparisons between the man-made and the biologically evolved.

For instance, it is often difficult to discover what a given structure "is for." I should not even want to know how many of my hours have been spent in pondering the direct adaptive significance of various bones, joints, tendons, and muscles, and whether the object of my obsession had a direct adaptive significance that could be intelligently pondered. It is likely that some animal structures could exist, not because they in themselves perform some function, but because they are developmentally tied to other structures which are useful [2]. So far there is no easy method for making incisive discernments in these questions. Too often the answers lie buried in the improbability of making a lucky observation on some aspect of the animal's life style. For years it seems to have been tacitly assumed that the precise abdominal bristle patterns seen in the many *Drosophila* species had no direct adaptive value. But careful observations showed that deviations from the normal bristle pattern, as could be studied in mutant flies, were adaptively unfortunate. Flies with too few bristles were prone to sink into soft, mushy substrates and become helplessly trapped, while mutant flies with too many would often draw beads of water from the same kinds of substrates, which would weigh them down and interfere with normal locomotion (Muller, 1950). Hence even a feature as apparently trivial as the body bristles of fruit flies are precisely organized in a directly adaptive manner.

The spectacular examples of adaptation readily seen in nature have had some bad effects on the mentalities of many biologists. It is incredibly alluring to believe that any organic feature has a direct, adaptive role, and most disastrously, it is incredibly easy to rationalize an adaptive explanation for very many organic structures. It is easy to fall into the error of posing elastic rationalizations, those sytems of "explanations" so meaninglessly broad and flexible that they can be fitted around all possible cases. If a portion of the skeleton is reduced in one species, its adaptive significance is the reduction of body weight; but if it is enlarged, it adds mechanical strength. This kind of simple approach leads us nowhere. It provides the means to gloss over some real problems on the slippery article of faith that whatever we observe must be adaptively perfect.

After all that, it must seem difficult indeed to reach some decision about the "perfection" of animal adaptations. It is. But we have some evidence bearing on the matter. Perhaps it will be only momentarily frustrating that not all the evidence is in agreement.

In Figure 20 I have a diagram of a python tooth. Teeth of this form are present on the paired maxillary and palatine bones of the upper jaw [3]. The frontmost teeth of these bones, especially of the maxilla, are mightily important in securing the prey. At a glance, these teeth appear to be very simple in form. Basically they are thorn-shaped, but a little closer inspection reveals additional details. The very tip of the tooth frequently shows a reversed curvature, and the shaft of the tooth bears one or more ridges which can, in some individuals, be surprisingly sharp-edged for such an inconspicuous feature.

These details are functionally sensible. The thorn-like curvature of the tooth is mechanically necessary to hold onto large prey. If the major part of the tooth were not curved backward toward the snake's gullet, a large prey

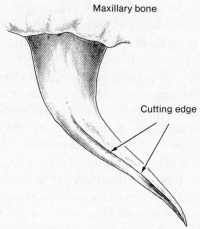

20   Python tooth, from the anterior end of the left maxillary bone.

that greatly separated the python's upper and lower jaws could pull itself free. But the backward slant of the tooth, although posing an obvious problem to the prey, poses one to the snake: the sharp tooth tips point in the wrong direction for the initial stabbing of the prey. This is lessened, however, by the reverse curvature of the tips. The forwardly turned tips catch in the prey's surface and greatly increase the chance that some of the rest of the tooth can then penetrate deeply. The cutting edges are situated so that, during swallowing after the prey has been killed by constriction, movements of the jaws will push these edges against the tissue in which the teeth are buried, enlarging the wounds and permitting easier withdrawal of the tooth.

These tiny features can thus be explained (hopefully not merely rationalized) on a functional, adaptive basis. So could the body bristles of fruit flies. Similarly, the genetic control of eye color in *Drosophila* seems to involve very precise mechanisms which confine phenotypic variation within remarkably narrow boundaries (Muller, 1950) [4]. The great precision and perfection of organic adaptations is suggested by other evidence, too, as in the phenomenon of "character displacement" (Brown and Wilson, 1956; but see Grant, 1972).

Character displacement involves two similar species whose populations partially overlap geographically. It is not infrequently seen that while animals from the nonoverlapping portions of the populations show few interspecific differences, the two species are readily discernible where their populations come together. In the regions of overlap, selection presumably favors individuals of one species who are different from those of the other. The effect is to perfect each species, according to its own fundamental features, in the utilization of limited resources, with the result that the species show greater differences where they are in competition. Often these differences seem trivial, and they can furnish some information on how small character differences reflect significant differences in adaptive style (i.e., niche utilization) [5].

Some of the most striking examples of precision must be those concerned with the form and coloration of many species which have undergone selection for close resemblance to portions of the environment or to other animal species. These resemblances, in the evolution of animal camouflage (see Portmann, 1959), or of mimicry (see Wickler, 1968) [6], can be almost uncanny. We have in these examples the advantage of knowing what the animal is resembling, what it is "supposed" to look like, and can measure the degree of conformity to determine levels of adaptive precision (although this has not, to my knowledge, been fully attempted).

These several considerations argue for precision and

perfection in adaptations. They indicate that animal adaptations possess some very narrow optimum form or expression, and that even slight deviation from the ideal can greatly diminish the useful performance of the feature. But there are other considerations, and these conflict with what has just been said.

The observed variation within all populations in nature certainly suggests a lack of adaptive precision by which such variation is tolerated. In some cases the variation of certain characters is discontinuous, and the population is thus *polymorphic* in respect to the feature. There thus frequently seem to exist in nature sets of phenotypic alternatives, of a continuous or discontinuous sort, which selection either tolerates as all being adaptively equivalent, or which selection is unable to reduce to a single, "optimum" type. Whichever is the case, there is implied a latitude of phenotypic form which is at least partially incongruous with the strictest notion of precision.

Some of the variation seen in populations is not due to genetic causes but results from external influences. This includes individuals who have suffered severe injury but who somehow manage to compete and survive alongside the undamaged phenotypes of their population. Some of these documented injuries reveal tremendous alterations in adaptive form. In most of these documentations there is no real information on the actual proportion of individuals surviving the injuries in the populations sampled. Almost surely such individuals are extremely rare in the majority of cases. The presumed rarity of these instances does not make them totally insignificant because, in evolution, new mutants might appear in the population as rare deviants from the normal pattern [7].

Not all individuals surviving injury are rare. Rand (1965) examined a collection of 164 individuals of a

Brazilian lizard species and found that about 12 percent of them had survived with significant injuries to the feet that resulted in the loss of at least one toe.

The most striking cases include injured marine fishes, some of which have suffered, but survived, a drastic alteration in body form. Figure 21 is taken from a paper by Gunter and Ward (1961), which provides additional examples. They point out that in some instances the manner of healing is most interesting, often involving compensatory developments of structures near the site of injury. For example, they show a croaker (Figure 21A here) that survived the loss of nearly all the caudal portion of its body. As the wound healed, the rays of the dorsal and anal fins grew backward, around the healing

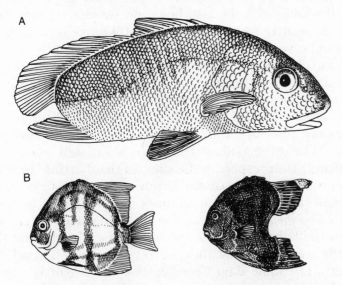

A. Croaker and B. spadefish, showing the ability of fishes to survive after major injuries. (From Gunter and Ward, 1961.) **21**

injury, and elongated as if to produce a crudely fashioned substitute for the missing tail fin.

Another factor that seems to raise a doubt about the precision of adaptations is the success of certain species when suddenly introduced into new environments. The number of successful artificial introductions is considerable and challenges notions of adaptive perfection. Thus, repeatedly, animals that evolved for great stretches of geologic time in one place, long selected for success in dealing with the physical and biotic features of that environment, are carried by man to a very different part of the world, where the introduced species instantly finds a place in the new environment and not infrequently competitively threatens native species who "should have" evolved adaptations so perfect for their own circumstances that no biological interloper could displace them. In this case as in the last, however, we lack reliable data on the relative frequency of successful introductions: they might be very rare events.

Finally, there are cases where certain structures in animals, although prominent, seem to have no direct adaptive meaning whatsoever. In pythons the premaxillary bone, the most forwardly situated bone of the snout, typically possesses four tooth-attachment sites, two per side. The premaxillary teeth are alternately shed and replaced with respect to the sites, so that barring unusual events, there is at least one firmly attached tooth on each side of the premaxilla at all times. In one type of python, *Chondropython* (which was discussed in Chapter 1), the premaxilla bears but one tooth site on each side, and the teeth are so tiny that they scarcely protrude beyond the gums. On some days *Chondropython* has a pair of seemingly useless teeth, and on others, because of the tooth-replacement pattern, there may be no teeth at all, or just one on one side. I cannot bring myself to believe that the premaxillary teeth of *Chondropython* have a significant survival value as teeth. However, it is possible that the vestigial premaxillary dental region, during ontogeny, interacts with other structures in important ways to ensure their proper development. If so, then the premaxillary tooth region is a positive adaptation (but not in terms of dental function per se). Wright (1956) [8] speculates that truly useless, vestigial structures will be rapidly lost in evolution. Nevertheless, for a structure or region whose task is only involved in the coordination of developmental events during embryonic life, the life-long cyclic activity [9] of the *Chondropython* premaxilla seems more than I should expect in a very well adapted system. This apparent imperfection is quite considerable, and improved biological alternatives are easily imagined. There is, perhaps, therefore, a barrier relating to genetic control, or to epigenetic integration, which thwarts evolutionary processes in the improvement of this system [10].

None of this evidence bearing on precision, pro or con, is particularly powerful. Much is circumstantial and seems contradictory. The contradictions might be unsettling for the wrong reasons, or for lack of enough reason. For one thing we have not yet decided whether we should expect most animal adaptations to be alike in terms of precision. For another thing, we lack statistical data, and we cannot be sure if the circumstantial evidence on one side is not very unevenly matched with that of the other. For a final thing, our definition of precision lacks rigor.

Implicit in assessments of animal adaptations very often is a notion of "how good or bad" they are in terms of what they are evolved to do. It is hard enough to find unambiguous meaning in that notion when applied to man-made, engineered articles. These articles should be more readily understood and analyzable, but there is always the difficulty of finding a criterion by which judgments can hope to have any useful meaning. By example, perhaps very many people have felt satisfied that the internal combustion, reciprocating engine found in most modern automobiles is really highly reliable and very good, and that despite its variations (overhead cam, V design or straight, etc.), the basic concept would be hard to improve upon. Yet recently there are major innovations that may represent significant improvements (e.g., the Wankel rotary engine), and there are growing and responsible complaints about the environmental pollution caused by the engines in use today.

In engineered products, criteria based on strict economic success (according to whether consumers purchase them) have certain unsettling facets when analogy is made with organic machinery. For in the world of commerce, engineering achievements such as electric can openers, motorized wax paper dispensers, and other contrivances which are neither useful nor esthetic are economically patronized by much of the human populace, uncovering some confusion about technical relevance and human gullibility [11].

Although there is great risk in reasoning by analogy, analogies play a part, sometimes an unconscious part, in much of human perception, and are different from genuine examples only when we are quite sure of the genuineness of the examples (which is not often enough). This discussion will now turn toward the pursuit of some analogies between the evolution of organic systems and the "evolution" of artificial machinery. We shall, at the outset, require a definition of precision that will serve for this exercise [12].

A simple way to define precision is to note how much

a machine can be altered before the quality of its performance drops below a given level. Thus a machine so "perfect" according to its own fundamental design that its ability to perform is hurt by even the slightest change is extremely precise. Another machine, performing just as well as the first, would be relatively imprecise if the quality of its output were little affected by small changes in its components. I am illustrating the difference between two such machines in Figure 22. In the figure there is no indication of a definition of "perfection"; the idea of "precision" is held separate from it. There is, though, no difficulty in introducing a crude concept of perfection here, by noting that if an organic system could undergo gradual alteration that actually *improved* it, the unaltered system would clearly be relatively imperfect. The key word always is "relatively." It would be satisfying to define a system in terms of whether it is "really" good or bad and not just *better* than another alternative, but I doubt that we can ever do this.

It is possible to relate the pictorial expression of Figure 22 to a well-known biological model. The Sewall Wright adaptive landscape model (Wright, 1932) has found wide application in diverse biological problems,

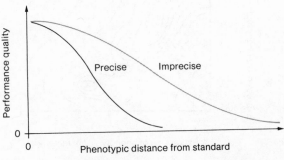

Relative adaptive precision of two hypothetical systems. **22**

and by altering a few of its original conventions, it can be employed here. A general representation of it is shown in Figure 23. Essentially it is a contour map. Convex contours ("mountains") are marked with plus signs, while concave contours ("valleys") are designated by minus signs. The contour lines indicate the height (or depth) and the steepness of the terrain (according to their proximity to one another).

The vertical and horizontal axes represent the degree of expression of two phenotypic characters that can be examined in the animal species in question. The heights of the peaks designate the degree of fitness of the individuals possessing a particular combination of character expressions as can be related to the two axes. For instance, in Figure 23 there are two peaks and one valley. The deepest portion of the valley corresponds to low expression of character 1 combined with moderately high expression of 2. Individuals possessing this combination are relatively very lacking in fitness, and selection will surely weed them out while preserving other kinds of

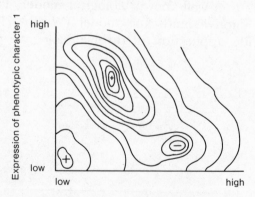

23 "Adaptive landscape." Explanation is in the text.

individuals. Contrastingly, individuals with low expressions of both characters are highly favored by selection, as shown by the peak in the lower left of the graph. However, the peak nearer the center is much higher, and individuals there have a greater potential fitness than those corresponding to the lower left peak. A change in the conventions to accommodate the terminology of Figure 22 would identify high peaks as places of high performance quality (instead of high fitness), and narrow, sharper peaks as specifying greater precision than wider, gentler ones.

If all the individuals of a population are identical, the population would be represented by a single point appropriately placed in the figure. On the other extreme, if the population were as variable as possible, with all manner of combined expressions of the two characters present in its members, it would not be a point but a square surface covering all the area between the axes. Generally, though, the variability of a population will lie between these extreme cases. We might imagine a population whose phenotypic range covered the topmost area of the lower left peak. Single individuals whose phenotypes lay outside that area, and thus had to be represented by points on the topography below the level of the maximum attainable at the peak, would be selected against. Any variation in phenotypes that places them down the side of the slope will be opposed by selection.

The lower left peak is not, of course, the highest in the adaptive topography. The peak above and to the right of it is significantly higher, and if it is unoccupied by a real population it would represent merely a potential that some population *might* attain if its members possessed the indicated phenotypes. Selection could not, even so, budge the lower left population toward the higher peak by gradual steps because the process would push the phenotypes into the valley between peaks, and this selection would vigorously oppose. In time, of course, environmental changes of all sorts will alter the adaptive topography, and could raise the intervening valley enough that an evolution from the lower to the higher peak could occur. This model, although conveniently expressed in terms of two phenotypic parameters, can be, and usually is, imagined to extend to a more realistic model containing many phenotypic traits where indefinite numbers of axes are envisioned. Let us hold on tightly to this model as we explore the patterns of change seen in many manufactured goods. In that, we have exceptionally good data on the evolution of the steam engine [13].

At the dawn of the eighteenth century, steam power was utilized by means of the Savery steam pump, whose task it was to remove water from mines. The device was simple, having a pipe extending down into the water which was to be pumped out, and a tank at the top. The tank connected with a steam pipe from a nearby boiler, and also with a water-ejection pipe. Steam would be allowed to enter the tank, where it was confined by valves that held it from escaping through the pipes. After the steam inlet valve was closed, the tank was cooled, producing a partial vacuum which, when the drawpipe valve was opened, sucked the water upward. This device was crude and did not produce sufficient pressure differentials to be very effective.

But in 1712 the Newcomen beam engine, which was totally superior to the Savery pump, made its appearance. The Newcomen engine possessed a cylindrical steam tank connected to the boiler, and a piston that was drawn downward when the steam was condensed by cooling.

The piston's movement activated a lever or beam which produced the pump stroke. All the movements were properly and automatically synchronized with the opening and closing of valves. This engine was admirably successful and was employed all over the world with little change from its original design.

Yet in 1769 John Smeaton made a thorough analysis of the Newcomen engine to discover that it was designed at great odds with optimal proportions. For instance, he found that the vacuum should be at 8 pounds per square inch for maximum thermal efficiency; when the pressure was either above or below this figure, the efficiency was lowered. Through relatively minor alterations, Smeaton thus greatly improved the original Newcomen engine. Clearly, Thomas Newcomen's engine, although successful, was relatively imprecise in myriad respects.

Soon after Smeaton's work there emerged a new and innovative engine invented by James Watt. His engine was very similar in appearance to the Newcomen type, but differed significantly in that the steam was drawn from the cylinder into a separate condenser. This disposed of the need to alternately heat and cool the cylinder, and was an important improvement [14]. It is unlikely, from all evidence available, that the early Watt engines were nearly as precise as the Smeaton–Newcomen device. Nevertheless, Watt's engine was a definite improvement. The engine enjoyed such commercial success that the firm of Boulton and Watt not only had trouble keeping construction up with demand but was too busy to prosecute most of the many imitators who probably were violating Watt's patent.

In Figure 24 I attempt to consider the adaptive values of three steam engines. The first "evolutionary" progression in this example is toward the very subtle refinements

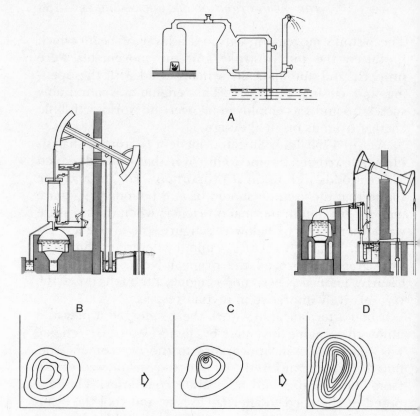

24 A. Savery steam pump. B. Newcomen beam engine. C. Smeaton improvement (not shown, except for "topographical landscape." In 1769 John Smeaton modified the basic Newcomen Engine, which little changed its outward form, but the subtle improvements were significant). D. Watt beam engine. The "evolutionary" changes in the beam-type steam engines are shown in relation to changes in relative precision as represented by the topography of "adaptive landscapes." Explanation is in the text. (From "The origins of the steam engine," by Eugene S. Ferguson. Copyright 1964 by Scientific American, Inc. All rights reserved.)

of a fundamental design, as in the superiority of the Smeaton type over the unmodified Newcomen engine. The second progression is not generated by subtle refinements of an old engine but by a basically new development in engine form and function; however, in this evolutionary transition there is a decline in precision despite the increase in performance quality (or fitness). The figure shows a wide tolerance of form for the Watt engine but an extremely restricted one for the Smeaton modification.

As subtle innovations appear, the Watt engine peak will become higher and narrower. When, however, a new innovation of a very basic sort comes into existence, it is likely that the new machinery will often be quite imprecise. This is a common pattern seen in many manufactured items, from timepieces to steam engines, firearms to fishing reels. Has this any meaningful comparability to biological evolution?

It is often observed in the fossil record that when an innovative adaptation emerges, it may appear with a bit of suddenness. Even if it is clear, and it often is, what sort of ancestral creature gave rise to the new form, there often are few or no intermediates preserved as fossils. The phenomenon is termed "quantum evolution" because, from superficial appearances, the evolution of the new form seems to have occurred not by a gradual process, but by a quantum jump [15].

The most reasonable explanation for this observation is that at the earliest manifestation of a new important adaptation, selection acts very rapidly to produce further evolutionary changes. This speedy voyage through the transitional steps has the implication of a relatively short transformation time and, consequently, relatively few intermediate forms. Intermediates are thus, quite simply, harder to find preserved as fossils.

Accepting this explanation brings a further possibility attached to it. The rapidity of quantum evolution could easily imply that the transitional populations were highly variable, offering selection a great deal of choice. A pattern such as this would be expected from the fossil record if new, large adaptations were initially imprecise in our terms, but alterable by continued selection, which, in promoting refinements through evolutionary change in the organic machinery, added precision. This is an interesting matter, for it touches upon a basic question relevant to Chapter Four concerning the evolution of novel adaptations: How sloppy can the initial steps be? The greater the sloppiness that can be suffered, the easier it is for a major change to evolve.

A second item related to the precision problem, but perhaps also touching gently our earlier abandoned discussion of perfection, is the Ludwig effect [16]. By this hypothesis a new phenotype could survive in a population even if it were inferior in the normal subniche of its species, provided that it could find a novel subniche of its own. Inferior types need only find a suitable subniche in order to survive, but of course it is that suitability which is the significant problem. When considering the Ludwig hypotheses we deal less with differences in performance quality between species and more with the difference between members of a population.

To an extent, any unusual phenotype will exploit its niche in an unusual manner. The very possession of changed adaptive equipment in an individual forces it to operate at least slightly differently in its environment. Whether the new subniche can support the novel phenotype is the crucial concern. There seem to me to be five basic factors that pertain.

1. The performance quality of the deviant phenotype in its subniche cannot be too poor.

2. The availability of the limiting resources in the environment must be sufficiently different in the two subniches. Suppose that an important factor is food and that the deviant type seeks prey in the trees while its conspecific neighbors feed on the ground. If tree prey never descend to the ground, there is clearly a private store of prey for the tree-climbing deviant. But if the prey moves freely from ground to tree, every prey item eaten on the ground is one less available in the trees. This latter case complicates things, and whether a basically inferior deviant could survive will then depend upon a number of details (e.g., the relative time that a prey individual stays in trees).

3. The difference between the deviant and normal types is important. If there is hardly any difference, the subniches will overlap to the extent that the inferiority of the deviant will make it a poor competitor in most situations.

4. It will also be important that the deviant does not adaptively approach individuals of species other than its own which share the same environment. Too great an adaptive resemblance will, again, lead to an overlapping of subniches, and unless the deviant is superior to the individuals of the other species involved, it will not likely survive.

5. It seems reasonable to expect that a large factor will be the relative abundance of the deviant in its population. If the normal type is very abundant, putting great competitive pressure on the normal subniche resources,

an inferior deviant type that was sufficiently rare might be favored in an otherwise unsuitable subniche, owing to the lower competitive pressure from the population at large.

I shall try to illustrate the last two points. Figure 25 is a very approximate first approach to the Ludwig argument. The abscissa represents subniche characteristics as if they graded evenly from one extreme (α) to the other (β) [17]. In this crude model we deal with two imaginery species, A and B, but our real concern is for A only. In Figure 25 I am showing curves of subniche quality for each species. Here I assume that in the subniche gradient there are, for each species, a range of subniches within which an individual can operate. Its survival prospects depend upon which subniche in the range it occupies. These survival prospects are based simply upon how well an individual of, say, species A, does in occupying a given subniche space when the number of other individuals in the same space is too low to significantly alter the intrinsic quality of the subniche. This is an attempt to evaluate the relative benefits of a subniche to an individual when considered quite apart from superimposing factors, such as the density of individuals in a given subniche.

As my diagram shows, the point of the subniche gradient at α offers the best intrinsic benefits for species A. To the right of that, the subniches are less good, and aside from distributional effects leading to densities in spaces beyond their capacities, the subniches to the right of α would decrease in value to each A population member occupying them.

There are two other curves drawn in Figure 25. Each one represents the negative number of individuals occupying points on the subniche gradient for the two species. The curves are drawn on the assumption that the

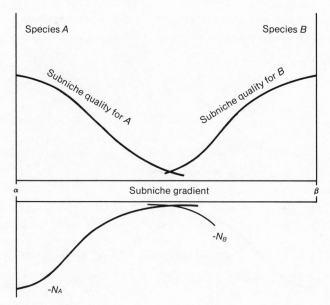

Implications of the Ludwig effect. Explanation is in the text.   **25**

greatest concentration of A individuals will be found close to α, where the subniche quality for survival is highest.

Figure 26 is the same as Figure 25, with the addition of some more lines. Here again we are concerned mostly with species A. If for each point on the gradient we make the simple arithmetic division of subniche quality (SQ) by the numbers of individuals (of both species—N_A and N_B), the generated curve $SQ/(N_A + N_B)$ gives us the ratio of subniche benefits to the numbers of individuals effectively lodged in a given subniche. These subniche benefits can be extended to many kinds of ecological terms, such as available food or space, protection from search patterns of predators, or whatever else.

The numbers of individuals at each gradient point thus becomes significant. If each point on the gradient

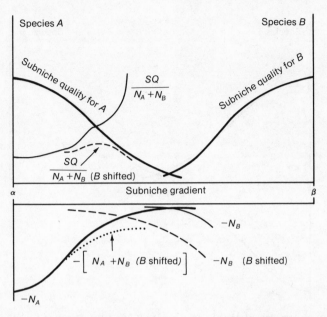

26   Implications of the Ludwig effect. (Compare with Figure 25.) Explanation is in the text.

represents a somewhat unique manner of dealing with the environment, and if each subniche point tends to be limited in terms of the density of individuals it can support, an advantage could be gained in occupying an empty subniche although it be intrinsically inferior to the normal subniche.

These are fairly direct extensions of the Ludwig hypothesis. In Figure 26 there are other extensions. Suppose that species *A* lives in a place where ecological "competitors" (however defined) are very plentiful, where the competition is "close" enough so that, in our simplified construction, species *B* is competitively nearer to *A*. Then the subniche gradient on which *A* could operate would be restricted. I have thus, on the lower portion of the figure, shifted the number of B individuals toward the left (as a dashed line). When the resulting division of SQ by $N_A + N_B$ (shifted) is drawn as a curve, the damping of the Ludwig effect by close competition is shown as the difference between the curves $SQ/(N_A + N_B)$ and $SQ/(N_A + N_B)$ (shifted). With sufficient scarcity of species where interactions between individuals most predominantly involve members of the same species, there is an increased likelihood that a novel phenotypic innovation with rather clumsy performance characteristics could persist. This potentiality is diminished when many species are present which can biologically encroach on deviant phenotypes.

Although there is nothing I can detect to be conspicuously unreasonable in all this discussion, we do lack concrete demonstrative evidence. I can, however, humbly offer one possible example, ringed, unfortunately, by mostly circumstantial evidence. It is based on two related species of very remarkable snakes.

If one looks at a map of the Indian Ocean, the eye can travel quickly from the east African coast to Madagascar, and then farther eastward, more slowly across the greater distance to the island of Mauritius (actually about 500 miles from Madagascar). It takes a very detailed map to show that several tiny islands lie near Mauritius, including Round Island, literally a dot in the ocean measuring but a mile across its largest diameter. Contained in the limited fauna of Round Island are the two snake species, related closely to one another and remotely to the pythons we considered in Chapter One. They may be termed bolyerine snakes [18], and their claim to distinction is a most unusual construction of the upper jaw apparatus.

Something is wrong with my drawing in Figure 27 if it does not show easily that the bolyerine has *two* separate

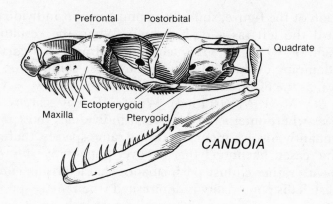

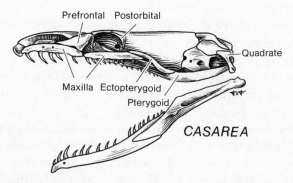

27 Skulls of the relatively typical boa-like snake, *Candoia,* and the peculiar bolyerine boa, *Casarea.* Note the divided maxillary in *Casarea.* (From Frazzetta, 1970.)

maxillary bones on each side. This feature is stunning. Nowhere in all the amniote vertebrates do we find a double maxillary bone. In bolyerines, the maxillary pieces are joined by a movable joint, and they are derived from a division of the single maxillary bone found in ancestral types.

Some very strange implications attach themselves to this condition. In the first place, an engineer would be more happily impressed with the mechanics of the bolyerine jaw than he would the more primitive condition. In bolyerines, there is a division of labor not found in their ancestors. The front part of the maxillary bone is free to tilt upward to a remarkable extent—to capture prey by a stabbing thrust—while the rear part, which is not important in prey capture, can move rectilinearly forward and backward. The rear part is thereby disassociated from a function it is never very good at in other primitive snakes, but in bolyerines its independence from the front part enhances its ability to pull subdued prey back into the gullet.

So then, why did no other python-like snakes achieve this jaw mechanism? This question impales us on the second point. How do you get a smooth transition from a single maxillary bone to a divided one? Either one has a fairly rigid single bone, or two bones connected by a movable joint. It is not easy to envision an adaptive intermediate between these two extreme conditions.

There is good reason to entertain the possibility that the first appearance of the joint in a population of protobolyerine snakes was very sudden. Either a pair of "normal" snake parents produced offspring who possessed this joint, or snakes with the disposition to snap the bone were favored by selection when the break healed into a joint rather than a rigid reunion of the broken pieces. If this latter possibility resembled the real circumstances, and if the double maxillary was an advantage in the early bolyerine's ecological surroundings, selection might have favored increasingly thinned, and breakable, maxillary bones until the thinness reached the point where the embryonic maxilla arose as two separate pieces [19]. In either case, the appearance of the joint would be sudden, a major deviation from the normal phenotype of the first bolyerine ancestors.

Although animals experiencing sudden changes in their adaptive form are buffered through compensations from related systems [20], it is hard to believe that the first bolyerine jaw with a divided maxilla was not a crude piece of organic machinery. Compared to the then-normal phenotypes, it had not undergone refining selection for generations and was probably extremely imprecise and imperfect. Yet, in an isolated environment, freed from the many competitively interacting species generally to be found on a mainland, phenotypic deviants could have had a chance to enjoy at least the mere toleration of selection. If the deviants could last in the population for enough generations, refining genetic combinations could be embodied in them, giving them a stronger advantage. In time, their advantage could become so superior that selection would alter the entire population in the direction of the divided maxilla.

Perhaps the initial step in bolyerine evolution was one of great imprecision but also of superiority (such as the Watt engine over Smeaton's). And perhaps the example has generality for the evolution of new, major adaptations. This must always be a matter of guess. Geological history is long; new, major adaptive advances have been relatively rare in the stream of time; and it is unlikely that we shall encounter such advances as first-hand, on-the-scene witnesses. Evolutionists concerned with the major alterations in adaptive form must always be interpretive historians. The fact that our present readings of evolutionary events, whether they be from laboratory or fossilized remains, have a weighty consistency is a hopeful indicator that we are on the track of understanding.

NOTES

1. The clock was made by Henri de Vic for Charles V of France (Milham, 1941).

2. I shall have more to say about this in the next chapters.

3. See Chapter 1 for a brief description of the jaw mechanism.

4. Muller's paper, "Evidence of the precision of genetic adaptation," is of great interest and importance, even though certain of Muller's interpretations of gene dosage effects are now viewed in a slightly altered light.

5. Hutchinson and MacArthur (1959) employ data on character displacement as a measure of the minimal adaptive differences between species which are necessary for coexistence.

6. See Wickler (1968). The adaptive basis of mimicry, where one species resembles another (usually a species that is recognized by potential enemies and, for various reasons, is avoided by them) is intricate and goes beyond the scope of this book. Wickler's clear, authoritative, and colorfully illustrated book should be consulted for further information.

7. The most remarkably deviant phenotypes seen in natural populations are usually of a nongenetic nature, as in the case of injured animals. Unusual phenotypes with a genetic basis seem much rarer. According to Devillers (1965), genetic changes that can produce large changes in one system often have general effects on the entire animal. The totality of the effects is so overwhelming that often the individual cannot survive.

8. Also see Van Valen (1960) for a discussion of "nonadaptive" evolution.

9. In all of this I am assuming a cyclic tooth-replacement mechanism similar to that seen in other snakes. Admittedly, no one has studied this process in the premaxilla of

Chondropython, although my examination of dried skulls in several museums has not revealed anything to disturb this assumption.

10. Bjorn Kurtén (1957) has investigated populations of fossil cave bears where intense selection for tooth characteristics failed to produce evolutionary change. He suggests either a developmental barrier or a lack of strong genetic control over the character being selected.

11. For a pertinent and sometimes hilarious comparison of good and bad engineering design, the book *Design for the Real World* by Victor Papanek (1972) will repay a reader's time.

12. The difficulties inherent in such an analogy should be clearly set forth. Some of the following have already been noted:

   a. The success of commercially made articles may depend more on public whim than on how well they perform their task.

   b. Machine "evolution" is often noncontinuous, occurring in "leaps" so immense that any comparison with nature is lost. There is little continuity between the horse-drawn carriage and the "horseless carriage."

   c. Basic machine parts, once developed, are often inserted in machines other than the one in which the part "originated." The sense of wholeness or continuity of the mechanical system undergoing innovative improvement can become very confused. (Mitochondria, chloroplasts, and stolen nematocysts might represent rare examples of this very thing in the biological world.)

   d. Often no effort is ever made, during any stage in the development of a machine, to determine its optimum form, and the degree of its precision cannot be appreciated.

   e. In both organic and inorganic machines there are types which, rather than performing a highly specialized function, are generalists in that they can deal with a range of related functions. Usually, however, they do not deal with any one function extremely well, although they may be marvelous at spanning a spectrum of functions with an *overall* adequacy and excellence. The definition of precision (as used in this chapter) should be applicable even in these cases, but will require a careful evaluation and definition of the machine's "task."

This list is not offered as a complete census of difficulties inhabiting the machine analogy. For further thoughts on the comparison between machines and animals, see Gould (1970). Rensch (1959) discusses convergence in biological form and technological innovations. He notes that the eyes of such diverse animals as Coelenterata, Annelida, Echinodermata, Onycophora, Mollusca, and Chordata are all of a type comparable in many individual details with cameras: "With regard to the convergent evolution of such eyes, it is remarkable that our instruments of photography are constructed in much the same way and contain the same essential parts (lens, 'retina,' pigment layer, 'accommodation mechanism,' iris diaphragm)" (p. 71).

13. See Calder (1968), Derry and Williams (1960), and Ferguson (1967).

14. Many other technical developments showed themselves in the early Watt engines. Among them was the Watt straight-line mechanism (see Note 1, Chapter 1), which, perhaps a little unexpectedly, Watt regarded as his most significant innovation.

15. See Simpson (1953) for examples and interpretations. It should not be thought that, while "quantum evolution" can be demonstrated in the fossil record, there are not many well-documented fossil cases of gradual evolutionary transitions.

16. Ludwig (1950). Also see Mayr (1963).

17. This is, of course, not very close to actual situations encountered in nature. There we could expect the subniche

parameters to be multidimensional, and some of them to be discontinuous and not behave as gradients. The diagram is, however, an easy way to come closer to an intricate problem.

18. For more on these snakes, see Frazzetta (1970).

19. Broken bones that fail to unite can produce a condition known as a "pseudoarthrosis." Literally taken, the term means "false joint," although anatomically it can be identical in form to a normal joint. A pseudoarthrosis often involves a reshaping of bone ends to produce joint surfaces and the development of synovial characters (see Murray, 1936). Pseudoarthroses are uncommonly noted in modern clinical records of human bone breaks (see Ham and Harris, 1971), although they have been an occasional medical problem in past years. In the possibility that bolyerines arose by selection for breaks and subsequent pseudoarthrosis formation of their maxillary bones, there is an almost unbelievable parallel reported by Ottow (1950; see also Hutchinson, 1962). Some of the large flightless (unfortunately now extinct) birds (solitaires) on Rodriguez Island (in the general vicinity of Mauritius) developed a weaponry of large, bony nodules beneath their wings. These were the results of bone breaks and subsequent healings, and Ottow postulates that the ability to experience such breaks and healings was under selective influence! In fact, the entire skeleton was affected, and the number of skeletal deformities observed in the remains of these creatures is very great.

20. There will be more about this in subsequent chapters.

6

TIME
AND CHANGE

For most of the five billion or so years of the Earth's existence, life, of at least a simple sort, has existed on the planet. Elementary organic things can be traced as far back as three or more billion years ago. It is not until the Cambrian period, however, that a great variety of complicated animals are to be found in the fossil record, and that time is a little over one-half billion years in the past. In fact, it seems to be true that there are no convincingly authenticated records of metazoan fossils until, or just before the Cambrian (Banks, 1970; Cloud, 1968; Schopf, 1970).

There must be little romance in anyone who can pick up an ancient fossil and not breathe a little faster. Here, in the hand, is a creature from times and places we can never fully sense. A thing that made its own short way when our distant ancestors were no more, but no less, than ponderous amphibians, or perhaps if just a bit further along, agile reptiles. A thing like a Devonian brachiopod, when discovered in its rock tomb and laid naked to an observer's eye, seems, despite its silence, to declare some wondrous message from the creatures of the past. That awesome brachiopod, or any once-alive thing that we might take in hand, is one of very many others that were not preserved for us. The probability that a long-dead animal will be left, more or less intact, must be imponderably low. It depends upon special conditions almost immediately following the animal's death and upon certain local geological conditions that must hold ever afterward. The probability also depends upon the makeup of the animal. Small, soft-bodied creatures have less of a chance than those with hard, durable skeletons.

It is the fossil record, almost alone, which is the convincing proof that evolution has occurred. There is an abundance of evidence from fossils that transformations have indeed taken place. Some are spectacular, others are minor. There are places where the paucity of fossils has left us guessing, and others where the record is complete enough to show the gradation of one major type into another, with a subtleness that makes the drawing of lines between groups a matter of pure opinion. There are breaks and discontinuities in the fossil record, but these are not surprising. It is far more surprising that the record is as complete as it is.

Fossils, their transitions, their dates of change— calculated now from techniques of atomic physics— are the firmest foundation we have for knowledge of evolutionary rates and of evolutionary directions. This great source of understanding, as consistent as is much of its ingredient information with other branches of evolutionary biology, can nevertheless offer itself to some bothersome interpretations. An intriguing one is summarized by James Brough (1958).

He calls attention to the time of appearance of animal fossils, beginning with the Cambrian. Certain types of animals, because of their bodily structure, do not fossilize well and must be pretty much discounted from the survey. These include some Protozoa, and the Platyhelminthes and Nematoda.

Beginning with the Cambrian, there are to be found

representatives of the phyla Protozoa, Porifera, Coelenterata, Annelida, Arthropoda, Mollusca, Echinodermata, and "Molluscoidea" (an inclusive taxonomic compartment containing the "minor phyla" such as Brachiopoda, Phoronida, and Bryozoa). Striking is the fact that with one exception, the phylum Chordata, all fossilizable metazoan phyla make their first historical debut in the Cambrian (chordates appear in the record shortly after the Cambrian, in the Ordovician). Furthermore, because the earliest Chordata were probably soft-bodied and not easily preserved as fossils, it is at least possible that they, too, existed in the Cambrian [1]. If that is true, no new animals, with an anatomical organization so different from others that they could be classified as new phyla, have evolved for the last one-half billion years.

As we descend to lower taxonomic levels, from phyla to classes, to orders, and so on, we are further enveloped by this curious matter. Major evolutionary events seem to have occurred long ago, with fewer such events as recent days are approached. In Figure 28 I have tried to summarize some of this information, but a little more should be said that can add a bit of detail which the figure cannot show.

At the level of classes, we obviously find in the Cambrian some of all those which will ever be represented as fossils. But following the Paleozoic era there are new classes arising in but a single phylum, the Chordata. These are the birds (Jurassic period) and Mammals (Triassic period) whose beginnings are to be seen in the Mesozoic era. In short, evolutionary novelties different enough to be recognized as distinct classes are scarcely to be found in the later times of metazoan evolution.

The pattern is repeated in the evolution of animals at the taxonomic level of order. Brough analyzed this by

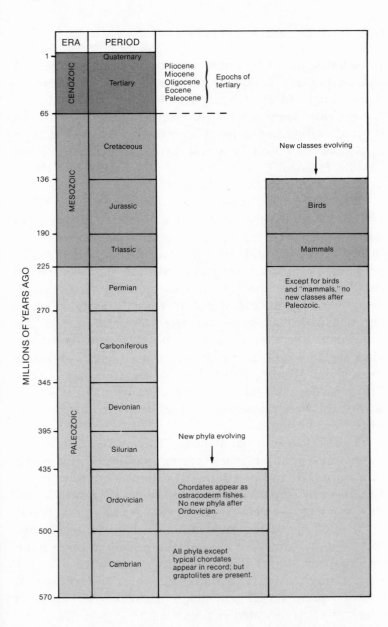

Geologic time scale. **28**

considering the orders of terrestrial phyla versus those of aquatic phyla. His aquatic phyla contain the Echinodermata, Coelenterata, Porifera, Mollusca, and "Molluscoidea," and they sum to forty-seven orders which have been identified in the fossil record. Of these, a full forty appear in the Lower Paleozoic, three more emerge in the Middle Paleozoic, and the remaining four reveal themselves sporadically through the Mesozoic.

The terrestrial groups Brough points to are the phylum Arthropoda (which include the Crustacea, an aquatic arthropod assemblage whose orders mostly arise in the Paleozoic) and the Chordata. In the Arthropoda, seven of the nine known fossil orders of spiders occur in the Paleozoic. In the Paleozoic we note the first appearance of sixteen orders of insects, thirteen more of which are newly displayed in the Mesozoic, and ten in the Cenozoic. The Chordata have no new aquatic orders after the Mesozoic, and only a few new terrestrial orders during the Mesozoic. But then, in the Cenozoic (in the Eocene), there is a burst of many new orders of placental mammals; but following the Eocene there are no new orders of chordates.

The conclusion that Brough reaches easily, based on these observations, is that evolution was more rapid in the past. It quickly produced greatly different adaptive types. But recently it has, in most cases, slowed down to the development of relatively trivial changes in animal form.

Brough speculates further that, when the whole history of metazoan life is viewed, there seem at times to be evolutionary surges in the sudden appearance of many new, though distinct, groups. He finds no correlation between these surges and major environmental changes which any sophisticated ecologist might expect.

Furthermore, Brough feels that many major evolutionary steps were not obviously adaptive. He speculates that it seems as if the major steps in evolution were not by Darwinian selection, but were perhaps due to higher mutation rates in earlier times which, almost, forced changes upon the animal populations of the long past.

I cannot accept all of Brough's conclusions, at least not as he states them, but his succinct statement of the apparent fact of early, rapid evolution bears on the problem of complex adaptations. There is here a pattern of major adaptive change which shows an unexpected concentration of evolutionary activity.

The pattern can to a less impressive degree be seen duplicated within a variety of animal groups at lower taxonomic levels. Simpson (1953) notes that usually a newly emerging group exhibits an early acceleration of evolutionary change which later slows, often dragging on at a low level for a very long time. The evolutionary schedule of change in lungfishes is not atypical. In Westoll's famous study (1949) of these fishes, a scoring system was developed which was modified by Simpson (1953) so that characters that were ancestral were assigned scores near zero, and modern, recently evolved characters were rated higher. The rate of change of the complex of all twenty-one characters regarded provides the pattern shown in Figure 29A. There is an early rise quickly climbing to a single peak, then a sharp decline to just above the zero level which is maintained for many millions of years. In Figure 29B the plot profiles the change of score with geological time, hence the schedule of "modernization" in lungfishes. The pattern can be appreciated in terms of rates of modernization of characters as in Westoll's work, or in rates of species production (speciation) within a group. A plot of the number of species members of a group frequently shows an early

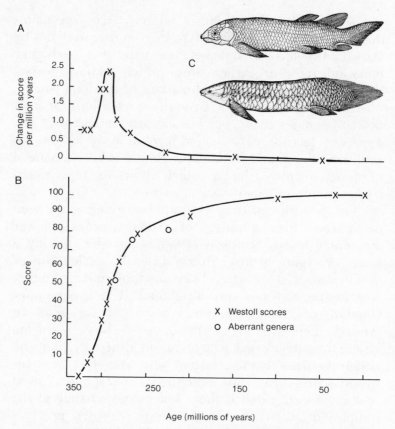

29 A. Evolutionary rate in lungfishes. B. "Modernization" of lungfish character complex. (Modified from Westoll, 1949, after Simpson, 1953.) C. Compares the oldest Devonian lungfish, *Dipterus* (upper drawing) with the modern Australian lungfish, *Epiceratodus*. (From Romer, 1970 and Norman, 1931.)

burst of new types, followed by much lowered rates of production.

Although this is a prevalent pattern, it certainly has its exceptions, as Simpson points out. Teleost fishes furnish one exceptional example, having not one but three speciation peaks in their history. Each peak, however, conforms to the early evolution of a distinct suborder of the teleosts, and if teleost fishes were divided into their lower taxonomic components, the single-peak concept would be upheld. To do this, of course, hints a little of cheating [2].

The lungfish example is interesting in some other ways. The group first emerges in the record during Devonian times and survives to the present as three genera. None of these is terribly different from the Devonian genus *Dipterus,* and I have in Figure 29C a comparison of this ancient form with the extant Australian lungfish *Epiceratodus.* The 300 or more million years of evolution seem to have wrought little in the way of startling changes. Other animal lineages also having their beginnings in the Devonian have not all imitated this sluggish evolutionary advance. One population of rhipidistian, crossopterygian fishes gave rise to the early amphibians, presumably within the Devonian period itself.

The generality of the evolutionary pattern showing an early rise in member species followed by sluggish evolutionary activity has been demonstrated in a fascinating analysis by Sloss (1950). In this work he produced frequency histograms of the time distributions of many groups of invertebrate fossils. Figure 30 shows two sample distributions from Sloss's paper. The histograms show the percentage of all species that exist in each of the indicated geological times. A quick glance at these might suggest that the number of existing species is greatest early in each group's history and then diminishes to remain at a relatively low level. However, Sloss recomputed his data as *cumulative* percentages with time—so

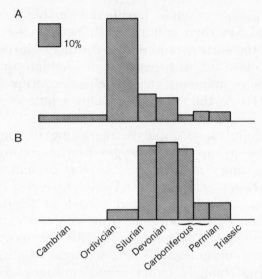

30 Frequency-distribution histograms of a fossil brachiopod group (A) and a group of corals (B). The plots show the percentage of all species that existed at the indicated geological time. (From Sloss, 1950; by permission of the Society of Economic Paleontologists and Mineralogists.)

that emergent species were represented as 0 percent at the very beginning of the evolutionary period, and as 100 percent at the end—and plotted the results on probability paper. Of the two coordinate axes of the probability paper, the ordinate represented a linear measure of time, and the abscissa, which is not a linear scale but arranged so that the intervals become abbreviated toward the 50 percent region, corresponded to the cumulative percentage. The plots were very nearly straight lines. This Sloss interpreted to mean that, in fact, the histograms such as I show in Figure 30 are not significantly different in shape than perfectly symmetrical, bell-shaped curves. He then performed mathematical differentiations, *d*(cumulative

%)/*dt*, on each cumulative percentage to produce new, theoretical plots of frequency distributions. A sampling of these is shown in Figure 31. The implied results are startling, for it is far from clear why evolving groups should display such a symmetrical distribution. As Sloss shows, the precision of this pattern must be due to a precision of origination and extinction rates, and these involve an early, rapid production of new species. This origination rate drops later, while the extinction rate rises.

The problem of the scheduling of evolutionary activity toward the beginning of a group's history, especially as

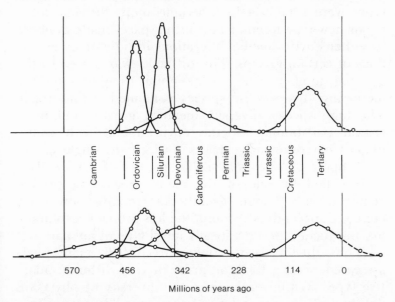

Frequency distributions of some fossil invertebrates as derived *31* by differentiation of cumulative percentages drawn on probability paper. Further explanation is in the text. (From Sloss, 1950; by permission of the Society of Economic Paleontologists and Mineralogists.)

it pertains to higher taxa, was examined by Valentine (1969). Essentially he showed a congruence of this pattern with good ecological theory. I have reproduced his diagram as Figure 32, which shows, in correspondence to Brough's data, the time of emergence of new major groups. Here there is also given the extinction rates at the indicated taxonomic levels. It is easily seen that higher taxa diversify earlier and that they also seem less prone to extinction than lower taxa. Valentine's reasoning is straightforward. In earlier evolutionary times the available niche spaces were filled by a rapid evolution that produced distinctly different adaptive types. Once these types were established, it became more difficult for a major new type to enter new niche spaces made available by either environmental alterations in time or by extinctions of certain groups. This difficulty was caused by the growing number of higher taxonomic categories and the increasing diversity of species contained within them. The high species diversity meant a greater and more critical partitioning of the environment, with consequently narrower niche spaces in which any single species could operate. A newly available space was far more likely to be filled by one of the many preexisting species, requiring only a slight evolutionary modification to occupy the space, than through the less probable evolutionary production of a drastically novel type of animal.

Thus, during early evolutionary history, many phyla appeared, setting forth an array of very different adaptive types. Within each, since the diversity of phyla was nearly sufficient to relate to most possible niches in a broad way, it was hard for evolution to "invent" another phylum whose adaptive strategy was not already represented by those in existence. But an increased diversity of classes could still occur. Then shortly later, when a

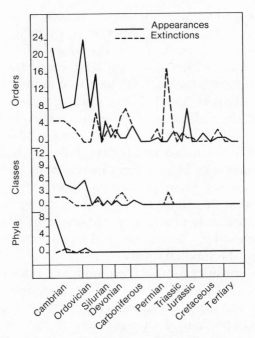

Appearances and extinctions of shallow-water, marine, and **32** benthonic (bottom-dwelling) invertebrates through geological time. (From Valentine, 1969; by permission of The Palaeontological Association.)

sufficient array of diverse classes existed, which could, through speciation within them, occupy most potential niches that would ever exist, further evolutionary activity could only be great below the class level. And so it was with successively lower taxa. The emergence of each evolutionary novelty at some given taxonomic level made it more difficult for further evolution at the same level [3].

Valentine's thoughts are interesting, not only in that his explanation is sensible in the present milieu of knowledge, but also in that there are noticeable implications for our concern with the emergence of major, complex adaptations. On the basis of his hypothesis, the wider niche breadth of the past must have eased the difficulty in selective tolerance of bold new mutations [4] (as discussed in Chapter Four). There must have been a greater likelihood that adaptive novelties of relatively imprecise and/or imperfect function could become established. The Ludwig effect, treated in Chapter Five, might have had an important role in evolution of the past. It does seem that the time of great evolutionary innovation is mostly over and that the diversity of living things we see today is due to a multitude of adaptive variations on relatively few basic themes. This hypothesis and its implications I try to support in the next chapters.

These considerations of evolutionary rates have been preoccupied with the *trends* of rate phenomena. There are numerous other evolutionary trends, but these mostly involve morphological changes that appear to be directional. For a number of psychological reasons trends are extremely difficult to talk objectively about. By necessity we view evolutionary history in retrospect and, knowing the ancestor and its remote descendants, can too easily project a directional pattern into the observations which they might not wholly deserve. There is, furthermore, an unconscious notion of evolutionary "purpose" which a few of my colleagues are prone to imagine. And I suppose this comes from the impression that some evolutionary changes look as if they had been "going somewhere" and resemble superficially the progressive alterations humans effect in their daily lives when they have a purposive goal in head.

I recognize that there do appear to be directional changes in some evolutionary transitions, but not for a moment can I accept the notion that, in some manner, the transformation is reaching toward some future state. Adaptations can only be evolved for the present condition, and there is no way short of magic to accumulate characters so that they will be on hand for some season yet to come.

One of the most discussed trends is the evolutionary size increase of many animals. There are many illustrative examples, including one given by Romer (1949) [5]. He notes that even in the evolution of horses, where there are sufficient known fossils to reveal that horse evolution was a complex branching and twigging of species in many adaptive directions, there is evidence of a size trend. Within the dense spray of phyletic lineages, the longest surviving lines show a progressive size increase. The many branches reversing the trend are short-lived. Perhaps this is a pattern common to a number of animal groups. It seems obvious that new groups did not mostly stem from the largest ancestors, for we do not live in a world of titans, but must usually have issued from modestly proportioned progenitors.

Within the broad category of "trends" is one of the most bizarre of all ideas relating to evolutionary changes. This is the concept of "racial senescence," in which it is assumed that as an animal group approaches extinction, it does so because it has become age-worn and incapable of responding appropriately in its evolutionary relationship with its environment. Adaptively it has become enfeebled and biologically bumbling. The evidence for this [6] is the "grotesque" evolutionary changes seen in animal groups just preceding their passing. Dinosaurs acquired exaggerated crests on their heads, titanothere mammals developed ludicrous horns on their snouts, ammonite cephalopods coiled and twisted in uncomfortable ways, and all this occurred just before extinction.

Quite aside from the comparison between "grotesque"

evolutionary modifications and the senescence of an individual seeming spurious to me, the grotesqueness itself might not even be true. Throughout much of their history dinosaurs have always been represented by unlikely looking beasts, titanothere horns seem no more preposterous than leglessness in snakes (which are not on the evolutionary decline toward extinction), and a recent study (Wiedmann, 1969) shows that the evolutionary alterations in ammonites during their final hour had been experienced by them before.

One of the most convincing studies of morphological trends was made by D. M. S. Watson on labyrinthodont amphibians (see Watson, 1951). This group of very primitive, often robust tetrapods, shows progressive changes in skull structure. Figure 33 shows drawings taken from Watson's book. The skulls are lined up (not precisely in a phylogenetic sequence) to illustrate the directional tendencies perceived by Watson. There are such changes as a broadening of the basal joint, enlargement of the suborbital vacuities, elongation of the parasphenoid, formation of the condyle as a double-knobbed structure, and flattening of the skull.

What is so arresting is that this trend occurred in not one but several distinct phyletic lines of labyrinthodonts, through roughly the same geological interval. More than that, Watson felt that this was not an example of adaptive responses in the different lineages to a similar environment; he looked upon the lineages as existing in different ecological circumstances. Watson's interpretation of this seems a little extreme, for he concludes that there must be an internal evolutionary drive within the amphibians, thrusting their phylogenetic changes toward an abstract goal that has nothing to do with adaptive correlations with the environment.

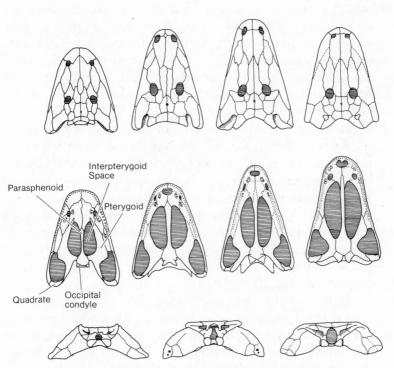

Evolutionary trends in the skulls of extinct, labyrinthodont **33**
amphibians. Top row, dorsal views, and second row, ventral
views of genera from left to right: *Eryops* (Lower Permian),
Rhinesuchus (Upper Permian), *Capitosaurus* (Lower Triassic),
and *Cyclotosaurus* (Upper Triassic). The bottom row shows
posterior aspects of the skulls of (from left to right) *Eryops,
Capitosaurus,* and *Cyclotosaurus.* The directional changes include
enlargement of the interpterygoid spaces, forward shifting of
the quadrates, the division of the occipital condyle into paired
bony knobs, and the flattening of the skull, among other
changes. (From Watson, 1951.)

The riddle of the amphibians was, possibly, partially solved through an analysis of adaptive trends in mammalian ancestors. The term "mammal" might at first seem to be pretty easily defined, for the diagnostic features [7] of "mammals" are totally unlike anything to be seen in other present-day animals. Years ago one could hardly have guessed that the fossil record would raise the possibility that a number of independent lineages independently acquired all or most of these characteristics.

Each of the "mammal" lineages arose from a therapsid reptile ancestor. The Therapsida was a group containing a number of different, but related types, and they all possessed some features reminiscent of the "mammalian" condition. A close analysis of them made several years ago showed that many "therapsids" had undoubtedly crossed the reptile–"mammal" boundary in terms of whole suites of morphological and physiological characteristics (Brink, 1956) [8].

Why should the specialized set of "mammalian" characters appear in several independent derivatives of ancestral therapsids? Olson (1959) compiled a long list of characters common to the early "mammals." He noted that the characters all appeared gradually in evolution, as modifications of preexisting structures. And he also noted that despite the close similarity of changes in these different lines, there was no niche correlation among them. Thus, despite the occupation of different kinds of ecological niches by the evolving therapsid lines, their characteristics changed in parallel directions at roughly the same time. The details here surely resemble those of Watson's labyrinthodonts.

Olson concluded that certain types of adaptations do not produce advantages that relate directly to the parameters of different niches, and are thus insensitive to their differences. Instead, these certain types of adaptations involve a general improvement which is advantageous in a variety of different niches. This general improvement takes the form of a modifying of preexisting structures and physiological traits, enhancing their general utility and thus, in effect, represents adaptations of the animal to itself. When certain of the therapsids began an alteration of their adaptations, the door was opened to selective favor for additional alterations of a complimentary sort.

Whether the kinds of adaptations in the labyrinthodonts can be explained in this way is not yet answerable. There is, however, an explanation that might touch the labyrinthodonts (and the late therapsids as well) but which is not much different from Olson's. In these animal groups, as in many others showing trends, the systems involved are highly integrated on a developmental basis, with built-in capacities and limitations involving more or less obligate correlations among their components. This kind of integration must place severe constraints on the kinds of changes that can occur. Certain changes could disrupt the developmental interactions of the embryonic building process or produce a functionally clumsy organic machine whose parts do not match sufficiently well for toleration by selection. It is not hard to imagine that in such cases, where the avenues of evolutionary change are few, somewhat *different* ecological modes could, through selection, produce *similar* kinds of changes. When dealing with related animals, as in the several labyrinthodont or late therapsid lines, we are involved with sets of animals having comparable integrative patterns. Hence, to overstate the case for emphasis, in some instances if the animals change at all, they will change in the same way.

I firmly suspect that these integrative patterns, which tend to be the relatively improbable adaptations spoken of in Chapter 4, are at the heart of both evolutionary potentials and limitations in much of animal evolution. I shall attempt to support this conjecture in the next two chapters.

NOTES

1. Graptolites, which are present in the Cambrian, are thought to be possible chordates.

2. Simpson (1953, p. 57) amplifies this point when he says that "the presence of more than one mode in a time-frequency curve, if based on adequate data, is a fact and reflects evolutionary events whether or not lesser, included groups are unimodal."

3. The convenience in saying such things as the "filling of niche spaces" should not convey the idea that there necessarily exists, at a given time and place, a definite number of niches. Whether a new species of animal can be added to an ecological community of species depends not only on the environment, on the number of different species already present, and on the abundance of species members, but also on the kind of species already present, the adaptive characteristics of the new species to be added, and on other factors. In general, however, the greater the diversity of pre-existing species, the harder it is to "place" a new type.

4. It is not, perhaps, insignificant that Brough considered mutations to have been of some unusual importance in the early history of evolution. In a sense he could well have been right.

5. But see Stanley (1973).

6. Some of it is nicely summarized in Rensch (1959).

7. These include hair, mammary glands, three ear ossicles, a single bone (dentary) in the lower jaw, a squamosal–dentary jaw joint, an enclosed braincase, secondary palate, heterodont dentition, and a diaphragm for breathing.

8. For some years following Brink's paper many paleontologists identified at least seven lines of "mammals" appearing independently during the Triassic. All but one or two died out sometime later (e.g., see Kermack, 1967). This view has been challenged by Hopson and Crompton (1969). They find that not all these Triassic lines conform to their stricter definition of a "mammal," and that the remaining lines are more closely related than used to be thought. However, we are still left with, at least, evidence of an approach *toward* mammal-like form by a number of independent therapsid lines.

7

THE GROWTH
OF FORM
AND VICE VERSA

My first evolutionary interests had a great deal to do with bones. I was drawn to them for the very unscientific reason that their variety of ornate surfaces, flowing twists and turns, odd-shaped apertures, hooks, and bumps made them, in my eyes, exquisitely fashioned works of sculpture. At the same time I had some other views of bones which, unfortunately, were less useful or truthful. I later had the task of ridding myself of notions that implicated bones as being rather inert, "stony," almost eternal.

Like other living systems, the vertebrate skeleton is capable of remarkable changes in growth and form which it undergoes throughout the life of its possessor. Here bone is a favorite subject of essays on the dynamics of growth. Part of the explanation for this favor is bone's capacity to become fossilized, permitting the skeletons of extinct animals to be studied. Another part of the explanation is that the hardness and durability of bone make for easy handling, storage, and measurement of bone specimens. And I suspect a third part is that most of us have had erroneous impressions about the inertness of bone in the predawn of our biological perceptions. One seldom succeeds perfectly in uprooting all traces of earlier prejudices, and the study of bone can often have some zest in uncovering a tiny, relict surprise.

Bone tissue itself has an extraordinarily precise structural organization of inorganic crystals (hydroxyapatite) and (organic) collagen, both of which are spatially arranged in specific patterns to form the substance of bone. Within the bone substance of "higher" vertebrates there are tiny spaces for living bone cells and for blood vessel channels. On a more macroscopic level a single bone may possess an internal marrow cavity lined with a thin sheath, the endosteum. Covering its outer surface is another sheath, the periosteum. Where two bones join to form a movable joint, the bone surfaces bear a layer of cartilage. If the joint can be properly classed as a true "diarthrotic" joint, there will be present a small "synovial" space between the articulating bones. The space is filled with a synovial fluid, a lubricant having some remarkable mechanical properties (see Alexander, 1968). A fibrous joint capsule runs from the nearby periosteal tissues, over the synovial region, and connects the two bones.

If a section is cut through a bone, as shown diagrammatically in Figure 34, it can be seen that some of the bone material appears dense and compact, while some, close to the marrow cavity, is formed as slender rods and wafer-thin ribbons and sheets. These slender structures are termed "trabeculae," comprising "trabecular bone"; the denser portions are "compact bone."

A given bone develops in embryology through one of several paths. It may begin taking form within a mass of fibrous connective tissue, as occurs in the development of superficial skull bones or of bony fish scales. In this case the finished product is generally termed a "dermal bone." But bones such as deeper skull bones, the ribs, limb bones, and portions of the vertebrae appear first as tiny cartilaginous replicas of the element to be formed. As

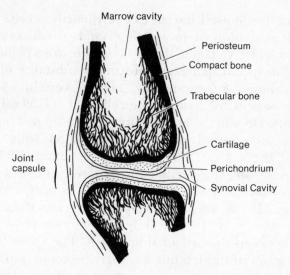

Marrow cavity

Periosteum

Compact bone

Trabecular bone

Cartilage

Perichondrium

Synovial Cavity

Joint capsule

34 Adjacent ends of two bones that form a diarthrotic joint, showing the internal bone structure and associated elements in a generalized vertebrate. The trabecular and inner surfaces (facing the marrow cavity) of the compact bone are covered by a thin, endosteal lining (which is not distinguished from the bone surfaces in the drawing).

development proceeds, the cartilage is gradually eroded away; simultaneously, bone is substituted for the vanishing cartilage. This process produces an "endochondral bone," which, in several important details of its embryonic growth, is not unlike a dermal bone. In both types, enlargement of the individual bones may occur by a continual remodeling (especially in "higher" vertebrates), mediated by two classes of cells. One of these, the bone-forming cells (osteoblasts), adds new bone tissue to the developing structure. The other, bone-destroying cells (osteoclasts), removes bone material previously formed. The two cell types work together, in marvelously coordinated cooperation, to retain the developing structure's integrity of shape and organization as it increases in size, and as it alters its form, as it must, while assuming new functional dependencies through ontogeny [1].

Throughout life, the periosteal and endosteal sheaths remain as a potential source of both osteoblasts and osteoclasts. This permits bones to grow after birth or hatching, and is also the basis for changes in form that can and do occur throughout an individual's lifetime. Perhaps the most fascinating aspect of these changes in form is that they may represent mechanically correct responses to changing environments. A bone is shaped by genetic controlling factors, but also by the patterns of growth of surrounding tissues and by external stresses that act on the bone. Of course, all these factors imply genetic control since, in any final analysis, the genes make bones capable of responding to all these influences.

For any given species there is a "normal" or usual pattern of stresses acting on a bone during ontogeny. These may come from the pullings and bendings of neighboring tissues, the stresses produced by contracting muscles, or the weights and impacts that bones must normally bear. A growing bone removed from its normal surroundings and cultured in isolation attains a general shape very reminiscent of the bone it is supposed to be, but it lacks many structural details that would normally be present. Direct genetic control carries the bone only so far in its developmental progress and then abandons it to the indirect genetic control mediated by external stresses. Without this, the bone is unfinished (Murray, 1936; Murray and Smiles, 1965).

Why does direct genetic control not complete the process? Because no two individuals are exactly alike in weight, form, muscular activity, and experiences. The finishing of a bone's shape is indirect from the genetic standpoint, but direct from a functional standpoint. A bone's shape directly adjusts to surrounding tissues. It is directly fashioned to stresses of muscle pull, to available blood supply, and to individual weight and use of the bone. The joint surfaces can adjust to one another in subtle ways. Thus bones compensate for individual peculiarities: the skeleton is tuned to its surroundings.

The adjustments of bone do not end when the embryonic period is done. Especially in "higher" vertebrates, bone can respond to changing patterns of function throughout life. These changes are observable in the size and shape of regions where muscles attach, in the overall thickness of bones, and in the internal architecture of the bone material itself. Alterations in architecture are readily seen in the directions of the trabeculae, and a heap of experimental evidence shows that internal bone organization responds with astonishing precision to changing patterns of external stresses. This response tends toward aligning the trabeculae in the "right" directions to best strengthen the bone to resist the stresses that it commonly experiences. If the stress patterns change significantly, so do the trabecular patterns. As D'Arcy Thompson pointed out in his *On Growth and Form* (1942), one of the truly great masterworks of biology: "If a bone be broken and so repaired that its parts lie somewhat out of their former place, so that the pressure- and tension-lines have now a new distribution, before many weeks are over the trabecular system will be found to have been entirely remodelled, so as to fall into line with the new system of forces" [2].

Other adjustments in bone can be seen. If the tibia, the larger of the two bones in the shank of the leg, is removed, the slender fibula which remains will widen to enlarge its weight-bearing, cross-sectional area [3].

Many of the changes in bone are mediated directly or indirectly by agents causing mechanical stresses. Bone will respond directly to stresses produced by the action of muscles. But also the rich blood supply of muscles can, where the muscles attach to the bones, contribute aid to the formation of bony prominences (Boyd et al., 1967).

There is here in the bone–muscle system a grand adaptation for adaptability. An individual whose life experiences repeatedly involve certain kinds of muscular activities will increase the size of particular muscles, while the bones are being altered to better accommodate the particular life style. There will be concomitant increases in muscular blood supply, enlargement of connections at joints, a gradually revised "sense of balance," and other behavioral recognitions of the altered bodily structure. There also may be compensating shifts in other portions of the skeleton (Thompson, 1942).

In other organ systems, too, we see abundant evidence of basic adaptations for adaptability. Functional necessities tend to stimulate appropriate responses. In vertebrates a high-protein diet can stimulate enlargement of the kidneys, cardiac enlargement can result from hypertension, the adrenal cortex thickens under stressful conditions, and low calcium intake will be followed by parathyroid enlargement. If pieces of liver, salivary glands, kidney, and other organs are removed, the remaining tissues will enlarge to compensate for the loss (Goss, 1965). In certain aspects of their effects, these changes are not unlike those wrought by the physiological mechanisms for adaptability to changing conditions, such as for example, shivering mechanisms in mammals at low temperatures or the ability of superficial blood vessels to change diameter in response to temperature differences. There are analogous biochemical examples, such as the blood buffer system. The phenomenon of wound healing, even of psychological repair, are further types of adaptability. The mechanism of gene control (see Rendel, 1968) suggests adaptability. The capacity to learn, seen in many animals, belongs to this list of adaptability mechanisms. The developmental process itself, initiated by gene action and carried along by the interaction of gene products in a manner to ensure mutual adjustment of each new organ or part with each other one, is among the most striking examples.

Adaptability is one of the major adaptations of animals. It seems to touch most organ systems if not all of them. Without it, it is possible that, at least during embryonic growth, the developing parts would too seldom achieve a proper integration with each other. To the extent that this kind of adjustment is necessary for successful development, each system has perforce been selected for its integrative capacities.

The degree of adjustment capability differs from one group of animals to the next, from one organ system to another, and varies in time during the developmental process. Waddington (1957) has given a pictorial representation of the interactive nature of development. Figure 35 is from Waddington's book and shows a portion A of the fertilized egg. This portion is comprised of subportions C_1, C_2, and B. The subportion B is altered through developmental time but, up until some point in this time, is operationally isolated from C_1 and C_2. Meanwhile, both C subportions are altered but maintain a mutual influence upon one another during this period. At some intermediate point in time, the C and B portions begin to interact, exerting molding or regulating controls

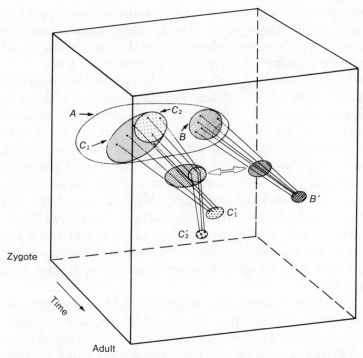

Zygote

Time

Adult

Representation of developmental changes occurring by **35**
epigenetic interaction and induction in the formation of a part
of an organic system. Explanation is in the text. (From Waddington, 1957.)

over each other (symbolized by the white, double arrow).
One result of this interaction is the strong alteration of
the developmental pathway of C_2 (to C_2'), whose direction
has shifted markedly.

This diagrammatic scheme suggests that beyond the
initiation of developmental processes by genes, there are
higher-level (epigenetic) coordinations of gene products
which evoke and guide further sequences in ontogenetic

change. The diagrammatic scheme finds easy support in experimental embryology. The molding action of an embryonic bone's surroundings is one of many examples of epigenetic action. Another is the evocation of major determinants of the fundamental body plan, such as the neural tube of chordates. For instance, as I show in Figure 36, embryos of all chordates possess a stiffened dorsal rod, the "notochord," running much of the length of the individual. It forms from the embryonic tissue layer, the "mesoderm," while just above it, a different tissue, the "ectoderm," rolls up in embryogenesis to produce the neural tube. In later life the notochord will have, in vertebrates, some role in the development of the vertebral column, while the neural tube will form the central nervous system. As different functionally as these two structures appear, they are related to each other in a vital way.

If in the embryo, the mesoderm (destined to become the notochord) is surgically removed, the overlying ectoderm will not roll up properly to form a neural tube. But if this excised mesoderm is transplanted into another embryo of the same age, by introducing the foreign mesoderm beneath the ectoderm of the host embryo's flank or belly, that ectoderm will form into a neural tube while the introduced mesoderm will develop as a notochord. Figure 36 shows an experimentally induced second neural tube in a host embryo.

These experiments demonstrate an interlocking dependency of one developmental process upon another. The processes manifesting themselves earlier in embryonic life are usually those of a most fundamental sort—they, the events they cause, and the further events caused, form a long chain of influences on the embryogenic schedule. An alteration of a "link" in this chain occurring earlier, rather than later, in the schedule will

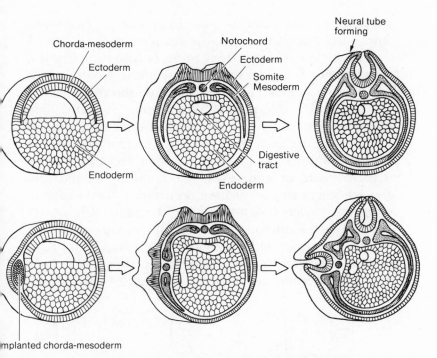

Sections cut from an idealized vertebrate embryo in late gas-   **36**
trula and neural groove stages. The top row shows normal
development, where the chorda–mesoderm forms the notochord and the overlying ectoderm rolls up to form a neural
tube. The bottom row shows the results of an experimental
manipulation, where a small piece of chorda–mesoderm is
transplanted beneath the ectoderm of the embryo's flank. The
transplanted tissue forms a second notochord (and associated
mesodermal structures) and induces the formation of a second
neural tube. Note also the secondary digestive tube.

have greater repercussions on the individual's development. The earlier developmental events thus tend to be
the hardest to alter without upsetting the individual's
viability and are often conserved throughout evolutionary

transitions. This must be the basis of the biogenetic "law"—that "ontogeny recapitulates phylogeny."

The recapitulation theory appears based on fact if a casual glance is flicked over what seems to be an endless list of examples. Aortic arch patterns, shown in Figure 37, have been interpreted as revealing recapitulation. The simpler primitive pattern, seen in some fishes, seems to appear in the early embryology of more evolutionarily advanced animals. In them, the aortic pattern undergoes change which, to some eyes, is reminiscent of the progressive stages that must have occurred in the lineage of ancestors. Other systems suggest the same sort of recapitulation. Embryonic changes in venous drainage, branchial pouches, kidneys, and somite musculature have all been involved in the recapitulation theory. De Beer (1958) has written an extensive criticism of this notion,

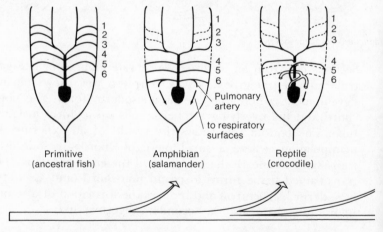

Primitive (ancestral fish)   Amphibian (salamander)   Reptile (crocodile)

Pulmonary artery

to respiratory surfaces

Phylogenetic history

37 Aortic arch patterns in several vertebrates, showing conditions in adults. Dotted lines show vessels that are present in the embryo but are lost later in the development of the individual. Recapitulation theory emphasizes that in embryology the and his book certainly should be read by those who find themselves either attracted or repelled by recapitulation. I find myself in considerable, although perhaps not total, agreement with De Beer. Examples such as aortic arches do not strongly suggest recapitulation to me. Instead they show simply how basic developmental formats tie in with the whole complex process of further ontogenetic change and thus tend to be adaptively indispensable. In terms of the aortic arches, an embryo amphibian has less to do with an adult piscine ancestor than it does with the embryonic stages of that ancestor. In both amphibian and fish the embryos have somewhat similar patterns. But in amphibians (and more advanced vertebrates) further modifications of that pattern occur. It is easier to see why certain embryonic patterns are conserved through evolution than why, in certain species, a fundamental pattern

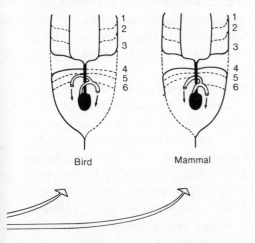

Bird   Mammal

primitive pattern appears but is later modified in a way resembling the phylogenetic changes through which a species' ancestors passed. Phylogenetic relationships of the groups are indicated. Further comment is in the text.

can be altered or "incomplete." These alterations do, however, occur [4] and tell us that as intricate as epigenetic dependency is, there are means by which the form of epigenetic relationships can be altered [5].

The interrelation of developing parts is, in some instances, exceedingly close indeed. This occurs especially where similar components form a functional pattern. In these situations, which are common, the structures are often difficult to consider individually. We can identify such patterns of growth in the bristles and eye spots (ocelli) in fruit flies, in the ankle and wrist bones of amphibians, and in the tooth rows of mammals. Surely such patterns exist in many more systems of many more animal species than have been studied so far. The foundations of such patterns are still matters of conjecture, but regardless of that, the subject belongs to the concerns of this chapter and cannot be excluded. It deals directly with integrated structural components and suggests, at least vaguely, how evolutionary changes that involve related characters may involve the overall pattern rather than its separate parts.

There are several general articles on the subject (e.g., Sondhi, 1963, and Van Valen, 1970) whose concepts and terminology disagree only slightly. In the development of patterns in the dental equipment of a mammal, or in the head structures (bristles and ocelli) of flies, there is hypothesized a reaction between two kinds of elements. One is a substance termed a "precursor" or "evocator." The other distributes itself unevenly across a physical substrate in greater and lesser pockets of concentration [6]. The distribution of the element on the substrate is termed the "prepattern," for it is the basis of the number and spatial arrangement of structural parts to be evoked by the precursor.

K. C. Sondhi (1963) performed selection experiments with a mutant strain of *Drosophila subobscura,* paying special attention to the head structures and comparing the flies produced in the experiments to wild-type examples. He noted that although the head structures, the ocelli and bristles, varied in number in the selected lines, the *pattern* of their arrangement did not. He concluded from this and other considerations that there existed an unvarying prepattern, specifying the placing of structures. But, in Sondhi's view, varying amounts of precursor in the selected flies determined the degree of head structure expression at the assigned places in the prepattern. Sondhi's diagrammatic model is reproduced here as my Figure 38.

Van Valen (1970) sees such a developmental scheme as explaining functional correlations among related parts of a system. Mammalian tooth batteries show such correlations, and the discovery by Kurtén (1963) of the loss, and subsequent return, of two portions of the lower molar tooth in fossil examples of the cat *Lynx* suggests a dental prepattern that, although not always expressed, was never lost.

The prepattern–precursor theory, if true in any essential respect, is significant not only in its demonstration of how functional components are orchestrated, but how evolutionary shifts can readjust all the parts into a new working relationship. The very presence of regulating factors that control the entire set of components could provide a potential array of somewhat different, viable editions of a single system, at times capable of accord with different paths of selection pressure.

In many cases the relationships among growing parts can be expressed in simple mathematical form. The fact that such relationships often seem quite precise, showing

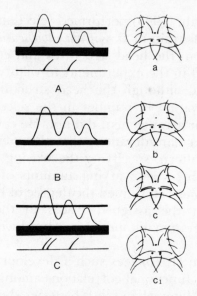

38 Heads of fruit flies, *Drosophila subobscura,* showing bristles and ocelli, and hypothetical diagrams of prepattern–precursor reactions. Head a is the wild type. Diagram A represents three peaks of the prepattern. The upper horizontal line represents the precursor, which, cutting across the two peaks as shown, evokes a structure (a bristle) at each. The third peak, although present, does not produce a structure because more precursor is required than is available. Head b has less than normal precursor, and heads c and c_1 have slightly more than normal amounts of precursor. (From Sondhi, 1963.)

firm statistical correlations, may be taken as additional evidence—if additional evidence is needed—of ontogenetic integration. Two parts of an animal can be measured and compared at different stages of ontogeny so that the size of one is quantitatively related to the size of the other. In some analyses the total length of the animal is taken as one of the "parts," and the size of a particular organ is compared to it. In others, two organs are compared. Or two different measurements of the same part (width against length for instance) are used.

What is usually found by this comparison is what should be expected. The relative growth of two parts is seldom linear. One measurement increases out of proportion to the other. As an animal grows, its linear dimensions, surface areas, and volumes would increase at different rates if a very young animal was a perfectly proportional replica in miniature of older growth stages. For functional reasons, therefore, nonlinear relative growth (termed "allometry") should be encountered more often than linear ("isometric") relationships.

A simple relative growth equation has the form [7]

$$y = bx^{\alpha}. \tag{1}$$

The terms y and x are the measurements being compared. The b and α are constants, and it has been many times stated that α contains the most "biological significance." As we shall see shortly, however, the b term possesses important implications of its own.

Equation (1) can be graphed, as I have done in Figure 39A. Here y is a measurement that increases disproportionately faster than x. This general sort of curve would be generated if, say, y were the diameter of a leg bone and x the total length of an animal. As the creature enlarges, its volume, and therefore its weight, tends to grow faster than the cross-sectional area of the supporting leg bones. Hence, to keep up with the rapidly increasing weight, the leg-bone diameter must grow faster than the animal's length.

In a graph of the form shown in Figure 39A it is not easy to discern the separate effects of the two constants b

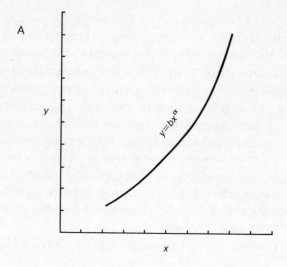

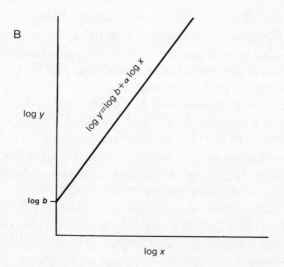

39 Graphs of the equation $y = bx^{\alpha}$ on the linear coordinates y and x, and of $\log y = \log b + \alpha \log x$ on a logarithmic scale.

and α. But equation (1) can easily be expressed in logarithmic form,

$$\log y = \log b + \alpha \log x. \qquad (2)$$

The form of this expression tells us simply that a plot of logarithms of y and x, rather than of y and x themselves, will provide us with a straight line that intercepts the $\log y$ axis at point $\log b$, and whose slope is α. Figure 39B shows it. The fact that the slope of the line reveals α makes the value of this term instantly perceivable. If y grows proportionately faster than x, the line will be tilted upward, making more than a 45° angle with the $\log x$ axis. If y grows proportionately slower, the angle will be less than 45 degrees. And if the slope should be exactly 45 degrees, $\alpha = 1$ and the growth is isometric, not allometric.

Among several examples I could choose to exemplify allometric growth I have chosen for Figure 40A an illustration from Huxley's classic book (1932) on the subject. The facial part of the baboon skull grows proportionately faster than the skull as a whole, producing a long-muzzled adult from a short-snouted infant.

A very significant feature of most relative growth phenomena is that the terms α and b tend to be constant through much of an individual's ontogeny, although there are instances where the value of one of the terms shifts rather suddenly from one value to another one. Frequently, when such a shift occurs, it is scheduled for the time when an individual faces a new adaptive path, such as at birth, or hatching, or molting, or metamorphosis. Even where shifts are shown to occur, a given set of values for α and b will usually hold for a goodly span of life.

A most remarkable aspect of the constancy of allomet-

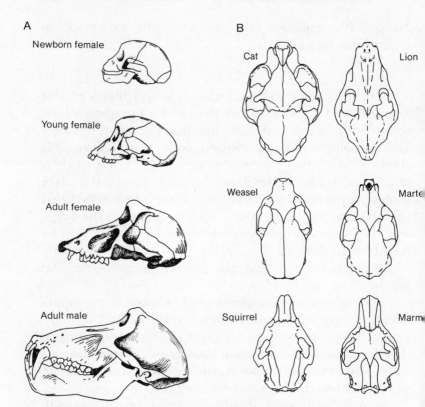

40 A. Series of growth stages of the baboon skull, showing positive allometric growth of the facial bones relative to the total skull length. (From Huxley, 1932.) B. Series of comparisons of facial proportions between two different, closely related species of unlike size (larger species on the right), to show conservation of allometric relationships in phylogeny. (From Rensch, 1959.)

ric growth terms can be made by comparisons of different, related species of animals. In examples of many such comparisons, the terms can be shown to be the same, or very nearly the same, in the compared species. There are certain allometric generalities that seem to hold even for rather distantly related types. In Figure 40B, for instance, pairs of related mammals are shown. One species of each pair is larger than the other, and a comparison of pair members shows that in phylogeny, as in ontogeny, the facial part of the skull often grows faster than the skull's remainder. The value of α in these examples, where the muzzle length is plotted against total skull length, remains strongly positive throughout many sorts of mammals.

The focus on the term α has often left the biological significance of b virtually ignored. But as Gould (1971) has recently emphasized, b can change to reduce (or enhance) excessive developments of certain features in phylogeny.

The adaptive utility of phylogenetic alteration of b is illustrated in Figure 41. If we imagine a species A as directly ancestral to species B, where the terms α and b for the allometric relationship between part y and total length x have remained unchanged, we can investigate the proportions of the two species at comparable stages. I have drawn dashed rectangles in the graph in the hope of making the comparisons more obvious. The larger species B has a greater y per x at any stage compared to species A. Clearly the two animals are not only different in size but in proportion as well. In some instances the excessive development of a structure in a larger animal species could be adaptively difficult. It might, in some cases, be advantageous, therefore, to modify the growth process such that the larger animal retained the proportions of the smaller one. Gould has shown that a change in b alone can bring this about, and will enable the successive growth stages of an animal to resemble in shape those of a smaller (or larger) species [8]. Thus b is a scale factor which, biologically, represents a mode of form determination which, mathematically at least, is

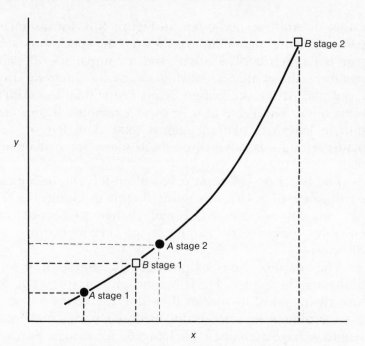

41   Hypothetical linear plot of part *y* against total length *(x)* for two
closely related but different-sized species *A* and *B*. An average
individual of *B* is consistently twice the size of an average *A*.
The values α and *b* are identical for the species, and hence life
"stages" for either can be found on the curve. The proportions
of the species for any given life stage are different, however, as
can be seen from the different shapes of the dashed rectangles.
For example, compare *A* stage two against *B* stage two.

independent of the factor α. Gould (1971) gives examples
of animal phylogenies in which *b* has been changed to
retain similar shapes among different-sized animals.

The allometric growth formula is a tremendously
simplified abstraction of an intricate developmental process. It does not reveal the machinations of development,
and it can only provide superficial clues about the nature

of the process. It shows what seems to be a conservatism—the stability of the value α—while allowing great shape changes within the unaltered context of the growth relationship. At the same time, changes in the scaling factor suggest a great flexibility within the growth pattern through phylogeny, which, again, points to a conservatism in developmental programming that does not preclude evolutionary variations on a basic theme. Size, shape, or both are significantly modifiable within a single dynamic pattern of development, which can be the root of more than one adaptive design. A single growth program can be the source of a considerable adaptive diversity in animal form. But it can also contain elements that limit the range of adaptations to certain directions and restricted distances. One of the most interesting studies ever undertaken toward illumination of these conjectures is now very old.

In 1934 A. H. Hersh made an allometric growth analysis of an extinct mammal group, the titanotheres, whose members existed in the Eocene and Oligocene. Titanotheres were great in size and grotesque in appearance. Many of them sported massive horns at the front of the skull, and possessed a thick, clumsy-looking skeleton (Figure 42). Fortunately, they seem to have had easily preserved bones for many titanothere species are known.

Hersh used measurements from various parts of the skeletons, comparing different species. One of his comparisons involved taking the skull width (across the zygomatic arch) and the (basilar) skull length. He averaged the values for each separate species, and plotted for each a single point on a set of logarithmic coordinates. I show Hersh's graph as Figure 43. Significantly, it was possible for Hersh to connect species belonging to the same genus merely by adding some straight-line segments to his graph. Each line segment, connecting its contained

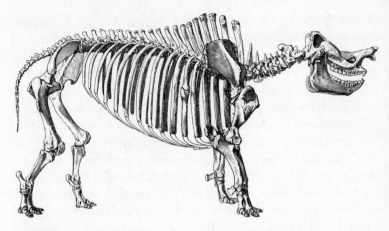

42 Skeleton of a large female titanothere. Males had different proportions and their horns were very much larger. (From Osborn, 1929.)

species points, was distinct from any other, and showed α and b values unique to a genus. I should think, if nothing else, that this result was a comfort to those who classify animals, who must decide which species of a major group are closely enough related to warrant inclusion within the same genus. But I think there must be much more to it than that.

The conservatism of α and b within a genus suggests a mode of integration involving skull width and length which is shared by the several contained species. In terms of these cranial features the "evolutionary distance"—however we might care to define that term—between congeneric species need not be very great at all; the *kind* of integration, at least insofar as α and b can reveal it, remains intact. But each basic integrative pattern can, in a sense, be pushed to varying degrees along one or more directions to produce different kinds of animals with somewhat different adaptive features. An evolutionary change in α is a much more major event, providing a new developmental basis for structural organization, and bestowing the capacity for a new set of variations on the revised theme. The sets of animals characterized by different αs and bs are, in fact, each unique enough to be recognized as distinct genera by taxonomic judgments whose basis for discerning relationships is not consciously concerned with mathematical constancy in growth parameters. The convergence of the

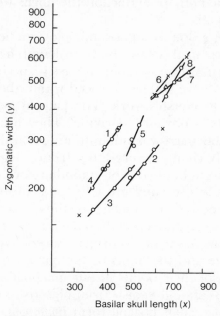

Double logarithmic plot of zygomatic width y against basilar skull length x for a number of titanothere species. Each point symbol represents averaged data for a single species. Species belonging to the same genus have their points connected by straight lines. The numbers within the body of the graph each conform to a different genus. (From Hersh, 1934; by permission of the University of Chicago.)

43

two approaches toward the same place cannot be idle coincidence.

A comparison between horn length and skull size led Hersh in 1934 to some unorthodox conclusions, which, in part, have been at least theoretically supported in 1971 (by Szarski). Eocene titanotheres had very small horns or no horns at all, and were generally of smaller total size than their relatives who were to appear in the Oligocene. The Oligocene beasts carried tremendous horns, and at the time of titanothere extinction the horns were half as long as the skull.

As before, a double logarithmic plot can be made for two characters, in this case the horn length against skull length. Now, however, a point for each specimen of all horned titanothere species is placed within the axes. Figure 44 is from Hersh's work. The pattern made by the plotted points is obviously striking. They nearly all fall within a notably narrow band, indicating that the α in this case has really changed very little (from a high value of roughly 9) throughout eons of titanothere evolution. The upper points in the band are Oligocene forms, and the lower points designate members of those Eocene species with horns barely large enough to measure or to estimate. If we turn now toward the hornless Eocene types, noting their skull size and the limits of the narrow band in the figure, and plot their points in our imagination, we find that the horns should vanish from objective human perception. In the small Eocene form *Eotitanops*, the calculated value of its horn would be ½ millimeter, utterly unfindable on its 313-millimeter skull.

From this it seems reasonable (even if odd) that long before titanotheres had visible horns, the developmental potentiality for them was present. The organization of skull ontogeny was such that as the skull components

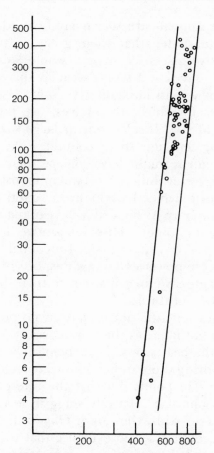

Double logarithmic plot of horn length y against basilar skull **44** length x. Each point designates a single *specimen*. (From Hersh, 1934; by permission of the University of Chicago.)

increased in size, one region would be influenced to enlarge much faster than the others. Then if selection acted strongly to favor increasing size through evolution, the horns would be brought along as a concomitant phenomenon.

As larger animals are selectively favored, there is of course no guarantee that their great horns would be happy adaptive assets. Perhaps it would sound partially convincing to say that it seems unlikely that selection for large size would exist immediately alongside the selection for great horns, which, themselves, are but allometric by-products of size. If selection for large size were sufficiently strong, enough so to counter whatever inadaptiveness the horns might have, the peculiar result could be a progressive evolutionary change, monstrously revising cranial form, which had no direct adaptive basis. This would be true only if the skull's developmental program were unable to loosen those integrative features symbolized by b and α without major functional damage, and if selection were powerless to favor aspects of the animal's structure or life style in a manner to turn the horns into an adaptively useful role.

In Hersh's view the horns were of no positive advantage at first, but later, as the horns became more prominent, since the beasts possessed them anyway, there was selection favoring those individuals who could make some use of them. He points out that the rate of titanothere evolution began at a relatively sluggish level but picked up perhaps tenfold by the Lower Oligocene [9].

There is, of course, no evidence that the horns were inadaptive until the Oligocene, and clearly some of the other views of Hersh are inferential at their best. The major question, hard to properly ask and harder to answer, is: How inflexible are the fundamental interrelationships of growth? If we view them as very malleable, we must expect that new integrative patterns can be evolved fairly rapidly. If we do not, Hersh's suggestion is not implausible. There does seem to be at least one example (in the teeth of extinct bears) where strong selection failed to produce evolutionary change (Kurtén, 1957).

Recently, Szarski (1971) has proposed that selection for one feature may produce change that exposes other features to new kinds of selection. Whenever the selection pressures on the various features can have the effect of enhancing each other, evolution could be rapid indeed. If Hersh was right about the titanotheres, once selection for large size exposed the enlarged horns to selection pressure, and once they had positive adaptive value, selection could proceed to increase body size rapidly *because* larger horns were advantageous.

There is one final study I shall mention in this connection. Like Hersh's, it is many years old (1928), and its conclusions quite unorthodox. The study was the product of G. K. Noble and M. E. Jaeckle, and it deals with the foot structure of tree-climbing frogs.

If one carefully examines the toes of tree frogs, there is found an expanded digital tip supported by an extra segment of cartilage between the terminal phalanx and the penultimate one. Terrestrial frogs lack the feature. Further examination of histological preparations sectionally cut through the toes reveals details of cell shape, arrangement, cell structure, distribution of skin glands, and arrangement of connective tissue fibers. In tree frogs these structural details seem reasonably correlated with their notable ability to adhere to a variety of vertical surfaces. According to Noble and Jaeckle, the histological specializations for climbing exist generally in tree frog species, even those not closely related to one another. On the other hand, terrestrial frogs lack the cartilaginous segment. They also show major differences from tree frogs in histological details of the toes; rather, most of them do. The exceptions among the terrestrial frogs, those that histologically resemble tree frogs, were given particular scrutiny. Noble and Jaeckle were convinced that these exceptional species could not be explained by supposing that their ancestors were arboreal. Nor could they note any pertinent functional differences in locomotion that distinguished them from most terrestrial frogs.

If their assessment of the phylogenetic relationships and functional characteristics of these frogs is correct (and it is probably not closed to challenge), it would appear that more than one type of complex, histological organization is allowable in the toes of terrestrial frogs. Perhaps the two structural versions are adaptively equivalent in surface-dwelling species [10]. Those terrestrial frogs resembling tree frogs in this manner might, however, be more probably the ancestors of future tree frogs than might other surface dwellers. Szarski's scheme would support such a view, and the thought was firmly in the minds of Noble and Jaeckle, for they say (p. 275): "The conclusion seems obvious that the tree-climbing apparatus developed before the frogs began to climb trees."

Such a conclusion should be handled with some caution, partially because of its sensitivity to underlying assumptions, and partially, unfortunately, because it is incongruous with the feelings of many evolutionists. It is, however, relatively more palatable in contrast with the surprising statement (p. 288) of the authors that a "detailed analysis of the many 'marvelous adaptations' in the Amphibia will reveal, we believe, that in *most cases* [my italics] the modification arose before the function." I personally doubt the truth of the statement as expressed. But when I reflect that G. Kingsley Noble was one of the greatest experts on the biology of the Amphibia who ever lived, I do pause.

NOTES

1. A third type of bone, exemplified by sesamoids and some other structures, is "metaplastic bone" (Haines, 1969). Unlike dermal or endochondral bone, metaplastic bone involves a change (not a replacement) of cartilaginous or tendinous tissues into bone. This appears to be a fairly recently evolved process, for no animal before the Jurassic shows evidence of metaplastic bone development.

2. Thompson (1942, p. 984). Thompson's book, published in its first edition in 1917, includes a basic treatise on the adaptive form of bone. The subject can be pursued through consultation of such important references as Thompson's book and that of Murray (1936). Other fundamental works are those of Bassett (1971), Currey (1968), and Enlow (1963).

3. It seems that this widening is not totally related causally to the relative increase in stress that the fibula must bear when the tibia is removed. Murray (1936) offers discussion.

4. There are many examples of species which, in regard to certain systems, display alterations of certain early embryonic patterns. In therian mammals there is often a tendency for aortic arches 1 and 2 to appear and disappear sequentially, so that the two exist at different times in a manner unlike the case in many other vertebrates. It is often noted that arch 5 fails to show itself at all in human embryology.

5. I shall return to this briefly in Chapter 8.

6. That a substance upon a substrate will form uneven, unvarying concentrations is given by the "Turing hypothesis." According to it, the concentration patterns of the substance will form "standing waves" of a predictable kind. A simplified, somewhat intuitive approach to the Turing hypothesis is given in J. Maynard Smith's book, *Mathematical Ideas in Biology* (1968b).

7. Two recent, general works on the subject are those of Alexander (1971) and of Gould (1966).

8. That a single value change of b can shift the entire set of growth sequences to correspond to larger or smaller sizes was shown by Gould (1971). If α remains unchanged, a change in b that makes two different-sized species identical in shape at an equivalent life stage will also guarantee their identity in shape at all other comparable life stages. Gould illustrates this by a logarithmic model, but I shall offer here a linear version. We want to show that when species B averages n times larger than A in any given life stage, a single value of b can be found which will specify that A and B will have similar shapes throughout all comparable life stages.

For A we have at some early stage that $y = bx^{\alpha}$. As the animal grows it reaches a new growth stage when its size has increased to $x + \Delta x$. The size of the "y" organ is now $y + \Delta y = b(x + \Delta x)^{\alpha}$

B, however, is n times larger, and if A and B are assumed to be of similar shape, the equation for the early stage of B (comparable to that of A) is

$$ny = bh(nx)^{\alpha} = bhn^{\alpha}x^{\alpha}, \tag{1}$$

where h is a constant multiplier of b of such value as to satisfy the assumption of similar shape in A and B at *this* stage. At the later stage we find that the equation for B can be written

$$P(y + \Delta y) = bhn^{\alpha}(x + \Delta x)^{\alpha}, \tag{2}$$

where P is the size of the "y" organ in B, relative to that in A, at *this* life stage. We need now only show that $P = n$. Substituting $y = bx^{\alpha}$ into equation (1) we have

$$nbx^{\alpha} = bhn^{\alpha}x^{\alpha} \tag{3}$$

which reveals that $h = n^{1-\alpha}$. From this and from equation (2) we can write

$$P(y + \Delta y) = b(n^{1-\alpha})n^{\alpha}(x + \Delta x)^{\alpha}$$
$$= bn(x + \Delta x)^{\alpha}, \tag{4}$$

but substituting $y + \Delta y$ for $b(x + \Delta x)^{\alpha}$ on the right offers us

$$P(y + \Delta y) = n(y + \Delta y) \tag{5}$$

which simplifies to $P = n$.

9. Hersh used the published bone measurements of Osborn (1929) and he followed Osborn's interpretations of titanothere phylogeny. Van Valen has cautioned me that Osborn's analysis of Oligocene stratigraphy is possibly faulty, and any conclusions based on events within this epoch in titanothere evolution must be tentative.

10. Of interest is a paper by Hobart Smith and James List (1951) reporting on the presence of expanded digital tips in an aberrant specimen of the bullfrog *(Rana catesbeina)*, certainly not a tree climber. The superficial resemblance to tree frogs did not include the presence of the cartilaginous segment.

8

~~~~~~~~~~~~~~~~~~~~~~~~~

EVOLUTIONARY
ANTICIPATION
FROM
ADAPTABILITY

"Preadaptation" is a treacherous term. It has too often kept intimate company with some very naive evolutionary thinking. Especially in the older literature can be found "examples" of animal structures which, when they first took their place in evolutionary history, had no purpose other than to wait to become an important adaptive feature of a remote descendant species. Whenever these examples got serious attention it would be revealed that the regarded structure *was* an advantage to its earliest possessor or was correlated with an advantageous feature. The association between the word "preadaptation" and dubious teleology still lingers, and I can often produce a wave of nausea in some evolutionary biologists when I use the word unless I am quick to say what I mean by it.

What I do mean by it is that when a new feature spreads through a population it does so only because it possesses a selective advantage at the time [1]; *but,* in certain instances, it happens that a descendant species under different selection pressure will adaptively utilize the same feature in a new manner. In the early evolution of vertebrates, fishes evolved a pair of pectoral and a pair of pelvic fins. This was an adaptation for improved efficiency in swimming. It had not a thing to do with the convenience of two pairs of limbs for life on land for the first terrestrial vertebrates, which were, in time to come, to issue from one of the fish populations. The availability of two pairs of limbs to the emerging land forms was mere serendipity. If no fish ever possessed paired fins, vertebrates might never have got out of the water, or if they did, they might have utilized a locomotion method that in structure and function would seem pretty queer to us as we are now. The chances that we would be as we are now, however, would not be overwhelming.

Thus, in a very major sense, preadaptation refers only to the dumb accident that a structure adapted for one thing now can be adapted for something else later. But this can be seen from another angle. Since at least most animal structures are derived from preceding ones, most (or all) adaptive change is based on preadaptation. The latter view does not change the fact that we are still dealing with accidents, because adaptations that have occurred early in our ancestry were not the only ones that could have possessed preadaptive potentialities.

An adaptive system, evolved in an animal group at any point in time, will possess certain potentialities for further modifications. Within broad limits the evolved system will determine the direction of whatever evolutionary change might occur in the future. This much must seem obvious. I hope that it will also become appreciated that certain adaptive features not only set possible directions for future evolution, but can tend to promote future evolutionary change itself.

This would occur whenever an adaptation possesses the ability to "anticipate" selection pressures yet to come. When I say that selection is often likely to favor structures that are not only currently adaptive but that will anticipate adaptive needs of the future, I have not slid into a teleological revery. In the first place, there are real examples of what can be called anticipatory adaptations. The abilities of insects to quickly adapt to insecticides that are unlike toxic agents encountered in the past; of microorganisms to become resistant to new antibiotics; of vertebrate immune systems to effectively handle new substances; of birds in England to be capable of learning to open, and consume, the contents of milk bottles left on doorsteps; of the New Zealand kea, a parrot-like bird, to acquire the habit of attacking and devouring the introduced sheep, are all suggestive of some sort of anticipatory mechanism. There are other less impressive examples, and I believe the phenomenon has general importance. In a weaker sense, the ability of a new type of creature to author a radiation of many species hints at an anticipatory quality of the pertinent adaptive systems. We should need to ask how such a mechanism can evolve, and the question has already been nicely phrased by Melvin Cohn (1970, p. 255): "What are the common denominators of systems, present in an individual, that enable him to react *specifically* to a vast number of stimuli, unlikely to have been selective forces in evolution? For example, you can make antibody to D-tyrosine or learn to whistle Yankee Doodle."

If there is an answer available now, I think it must come from further consideration of integrative patterns in biological systems. We did some of this in Chapter 7, but now we should perhaps add greater emphasis to the responses of such systems to selection and to the flexibility inherent in them.

One type of flexibility has the ironic effect of reducing phenotypic variations among individuals. The individual deviations in form, which create the variations, could be the result of a number of things gone wrong in the developmental process. A gene mutation, mechanical trauma suffered by the embryo, or unexpected temperature or humidity changes are all potentially capable of twisting the path of development toward an abnormal, possibly inadaptive, phenotype. But there is, in many animals, a countering mechanism which can recognize a twisted developmental path and which, by appropriate government over further ontogenetic events, can switch the progress of development back toward its normal destination. The principle has been elucidated by Waddington (1957), who has given it the name "canalization." It is as if in the unfolding of epigenetic interactions the course of development proceeds along a channel, like a river flowing through a valley. Obstructions placed in the river bed might tend to divert the water up the sides of the valley, but as long as the sides are high enough, or the obstructions small enough, the water once past the disturbed region will flow back into the normal course. The system's canalization is made greater by raising the valley walls. But a sufficiently great disturbance can force the river over the walls, to run in a new channel toward a new destination (altered phenotype). The canalized system in development gives a stability or homeostasis to the outcome of ontogenetic changes (see also Lerner, 1954, and Rendel, 1967).

Figure 45 is a very simplified diagram of a model to illustrate the workings of a canalized system. It shows two elements or conditions, A_p and B_p, formed in development, which, at some time, must interact to form the next stage, S. If the system is disturbed by a mutant gene or an external factor so that the path that would have led to A_p is changed, a canalized system will make appropriate corrections. This may occur by directing development back to the proper "A" path after the disturbance. Or the path to B_p may respond to the deviating direction of the A_p-forming process in a manner to ensure that the normal S stage is still produced.

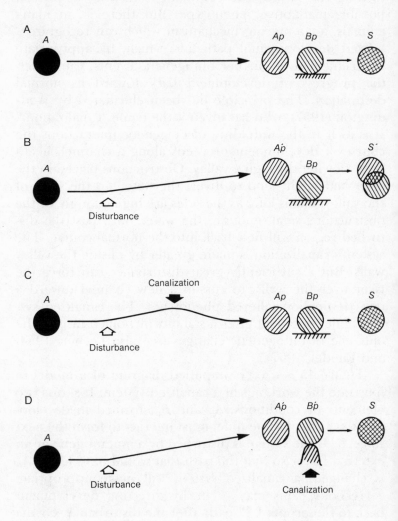

From an adaptive standpoint there should be no particular reason why a canalized response to a disturbance must result in exactly the same kind of structure that would normally be formed. It should be sufficient only that canalization furnishes the individual with an adaptation as selectively advantageous as the normal one [2], even if the canalized product is different from it. Waddington (1957) discusses an example: In mice, as in other mammals, the first two vertebrae behind the skull (the atlas and axis) are modified to allow a great range of head movements. The second vertebra, the axis, bears a bony, peg-like extension (the odontoid), which is loosely seated in the atlas just in front of it. The structure allows the atlas to swivel on the axis to permit the "no" gesture made by rotating the head. Ability to move the head in this fashion is important in all mammals, including mice without negative opinions. The absence of an odontoid in these mutant mice should be extremely inadaptive. In these mutants, however, there is a unique structural organization which permits individuals to make about the

Diagrammatic model of effects of canalization when the normal developmental process is mildly disturbed. A. Undisturbed: point A represents some condition which, in the normal process, is transformed to condition A_p. The latter interacts with B_p (which has developed from a preceding condition B) to produce the normal next stage, S. B. Disturbed, not canalized: a disturbance affects the "A" pathway to give abnormal A_p' (instead of A_p), which, in interaction with B_p, produces an abnormal stage S' (instead of S). C and D. Disturbed, canalized: when canalization operates, however, the normal pathway toward A_p is restored (C); or the disturbed "A" pathway evokes an appropriately altered "B" pathway (to give B_p' instead of B_p) (D). Normal development of S is achieved in both C and D.

same head movements as do normal mice. Instead of an odontoid, a special connecting ligament develops, and the joint surfaces of the atlas are changed. It seems that a canalization exists in the mutant mice, not to produce the same structure as in a normal type, but a different structure that functionally imitates the normal system.

Canalization could have been as easily treated in Chapter Seven as here, for the simple reason that it is but another example of the integrated multicomponent systems possessed by animals. Evolution's most major achievements are in this realm, and less so in producing—largely *from* such integrated systems—the small adaptive differences that have been the focus of most evolutionary research. Stebbins, in his analysis of progressive changes in evolution (1968; also see 1969), feels that the internal, genotypic environment exerts greater selective pressure upon new mutations than does the external environment. Whyte (1964, 1965) emphasizes the selection for internal compatibility in the individual's development.

These kinds of internal integrations all reflect canalizing phenomena in the broadest sense. The primary feature is the triggering of appropriate corrections by stimuli affecting one or more of a system's components. These kinds of phenomena certainly include the "Baldwin effect." Baldwin-effect examples have sometimes been displayed as if they were quirks of nature, understandable and compatible with selection theory, but of little general importance. Regarding this last point, I am not so sure.

The Baldwin effect is sometimes given the Lamarckian-toned misnomer, "the genetic assimilation of acquired characters." This unfortunate but superficially descriptive term has been applied because, in the Baldwin effect, one appears to see characters that once were acquired through use in an individual's lifetime but later seem to get programmed into the genome, for the characters appear at birth in new individuals. There are careful laboratory experiments which confirm the occurrence of this strange phenomenon (e.g., Waddington, 1953). Although laboratory experiments are the best real evidence we have of the occurrence of the Baldwin effect, I shall illustrate it by an example whose evidence is merely circumstantial. But the example is convenient since it is about a structure in mammals and easy to identify in ourselves. It concerns the callouses on the soles of the feet. These are developed by direct contact with the ground as the animal walks, and the more it walks the thicker, and hence more protective, the callouses become. Of course, callouses do not arise just anywhere on the feet; their distribution is directly related to which parts of the sole get the most abrasive stimulus, which itself is directly related to the shape of the foot and the usual manner of placing the foot on the ground. For different sorts of mammal feet there are different sorts of callous patterns. In many mammals, however, the embryo develops patterns of thickened areas on the soles before, obviously, it ever starts to walk. Thus a character that is evoked by direct environment stimulation is "assimilated" into the genotype in such a way that, now, the genes, or the developmental processes they govern, induce the character without need of the environmental trigger.

The secret, if there is one, is that the effect which was "assimilated" by the genome has always been there. The genes have, through a chain of interlocking processes, endowed the areas on the sole of the foot with the morphogenetic capacity to become thickened. In question only is what will set the process off. Will it be the abrasive stimulus from the outside? Or something from the inside? It is a bit misleading to pose the outside-stimulus question, for clearly the external abrasion does not, itself, produce a callous. Instead, it releases a series of related events, and one of *these* evokes the process. There is not one trigger, but a chain of them. To produce the Baldwin effect, selection need only eliminate the necessity for the first in the series of triggers.

It is not hard to invent a parade of evolutionary steps that could lead to the "assimilation" of foot callouses. In some hypothetical mammal species the members frisk about, scuffing their feet, and developing protective callouses. Those that can develop callouses fastest, in the right patterns, have a selective advantage over the others. Thus in the series of triggers each one is made increasingly sensitive, and less external abrasion is needed to initiate the process. At last a state of sensitivity is reached where the process is set off by such subtle stimuli that the need for abrasion is effectively eradicated.

The Baldwin effect, in particular the experimental evidence for it, suggests that once the complex mechanism for morphogenetic potentiality exists, minor evolutionary changes can mold its response capabilities along certain directions. The morphogenetic mechanism thus possesses some flexibility in how the process can be initiated, which is available to the grasp of selection. In the succession of ontogenetic changes which, compared among different sorts of animals, has led some evolutionists to speak of the recapitulation of phylogenies [3], embryonic stages are often missing or abbreviated in certain species. Frequently, these early stages function largely as epigenetic "triggers" that evoke later stages. But it is possible that certain of the vital responses can have their sensitivities raised by selection. Where this can occur, the "triggering" or evoking event may be reduced toward the invisible. The Baldwin effect, or a similar sort of thing, might be essential in explaining these "missing" ontogenetic stages.

A high degree of canalization or integration is potentially able to provide evolutionary opportunities in certain directions, but probably sets limits in other directions. A significant change in a complex system will depend upon major relational shifts of an improbable sort and will occur only as a comparatively rare evolutionary event [4]. In evolving a new system it would often be necessary to overcome the canalizing effect that maintains the balance of the old system, leaving the individual without the range of adaptability and integration possessed by its ancestors. This could easily be a selective disadvantage, if not immediately to one species, then at least to its descendants. Mayr (1963), Stebbins (1968), and Whyte (1964, 1965) all note that canalization and the possession of highly organized adaptive systems restrict the possible directions of evolutionary changes. The restriction is not absolute, of course. Complex systems can be revised in evolution, and when they are, a new, fundamentally distinct type of organism might be the eventual result.

Highly organized systems, because of their large evolutionary inertia, are likely conservative but, as I have been saying, probably underlie a great variety of adaptive forms. There are many suggestive examples. Jarvik (1959) examined some of the oldest fish fossils (ostracoderm fishes) and found that the body scales, beginning at the bases of the fins, ran farther out onto the fins and graded gently into the jointed, stiffened little rods that support the fins themselves. These rods or "fin rays" resemble those of living fishes, where embryological evidence shows that the rays are mesodermal in origin, developing in the dermis of the skin. The same is true of the body scales, however. Closer examination of fish embryology uncovers greater similarities. As Figure 46 shows, each portion of the fin ray begins development close to the interface of two tissue layers, the dermis and the more external layer, the epidermis. In development the fin ray begins its formation between the layers as successive generations of sheets of cells. Each of the first two generations of cell sheets sinks into the interior of the dermis, to there assume the structure of a definitive fin ray. But that is not all. The third generation of cell sheets lies close to the fin's base and becomes body scales. Over the entire fish's body, scales are formed by a peeling away of a cell sheet to become firmly associated with the dermis. And still that is not all. The peeling or delamination process is exactly what occurs in many parts of the internal skeleton. It can be demonstrated in sharks, for instance, that the skull is formed by successive delaminations of cells from near the dermis–epidermis interface and by the subsequent sinking of these cell sheets. The last cell sheet to delaminate forms the scales of the overlying skin. The basic delamination pattern is conservative, but the variety of structures it can produce is not.

There are other studies, such as that by Tarlo (1967), who observed mosaic scale patterns in ancient ostracoderm fishes, found functional meaning in them, and concluded that the skin of many vertebrates, mammals included, retains a basic mosaic pattern, again suggesting a conservatism in the developmental mechanism. An intriguing study by Dubrul and Laskin (1961) [5] shows how the simple fusion of two bones in the skull base of baby rats, when the fusion occurs abnormally early, can elicit a host of correlated skull changes. The modified

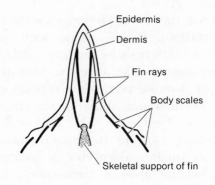

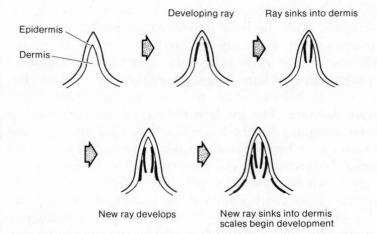

Diagrammatic section through a median fin of a fish to illus- **46** trate the role of "delamination" of tissues in the development of fin rays, scales, and other structures. Explanation is in the text.

skull resembles, in some ways, that of mammal species which, unlike rats, have an upright stance, and the authors conclude that some complex evolutionary changes in skulls have occurred without alteration of the basic growth patterns in ancestral types.

One of the most jarring examples is furnished by Hampé (1959). It has to do with the differences in hind-leg architecture in modern birds, in their now extinct *Archaeopteryx*-like ancestor, and in the reptilian ancestors of *Archaeopteryx*. Most reptiles possess, as we do, two bones (tibia and fibula) in the shank of the hind leg as shown in Figure 47. The tibia and fibula are nearly equal in length and in the ankle region articulate with the six or so small, cubical bones located there. As one approaches the bird-like *Archaeopteryx*, however, there are fusions and perhaps losses among the small ankle bones, leaving but two distinct elements, which, however, still articulate with the tibia and fibula. In modern birds the reduction of elements is carried even further; the changes in the ankle reflect the unusual mechanical requirements of a light-boned, bipedal creature which uses its legs to balance, leap into the air, and "catch itself" upon landing. The fibula is reduced to a small splint of bone lying against the top end of the tibia and does not extend anywhere near the ankle region. The two small ankle bones seen in *Archaeopteryx* are not free but are fused with the tibia. Hampé's technique was simple. He operated on chick embryos in a manner to keep the developing tibia and fibula from competing with one another for neighboring cells. When the fibula was provided a larger supply of cells than normal, it grew to a length fully as great as the tibia. But there was an additional effect. The two ankle bones grew separate from and unfused to the tibia. The resulting pattern was astonishingly reminiscent of the *Archaeopteryx* condition. Despite the marked differences between the ankles of *Archaeopteryx*-like creatures and birds, the morphogenetic relationships among the ankle components seem not greatly changed. One can imagine that a fairly minor

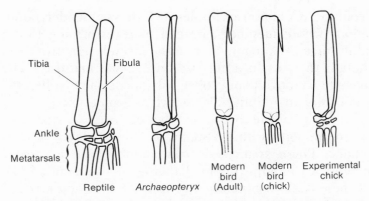

Tibia   Fibula

Ankle

Metatarsals

Reptile   *Archaeopteryx*   Modern bird (Adult)   Modern bird (chick)   Experimental chick

Leg and ankle bones of several animals. Explanation is in the **47** text.

gene alteration could perhaps control the amount of cellular material available to the fibula, but that this minor effect, when superimposed upon the system of integrative factors controlling ankle development, can and did result in a new functional, morphological construction.

Goldschmidt (1940) cited a number of examples of major evolutionary changes that could arise within the limits of a single type of morphogenetic system. His use of these examples to demonstrate large, macromutational alterations in form leaves me with the feeling that he often did not distinguish between truly innovative evolutionary steps, those involving revision of the interaction responses among components, and the flexibility of integrated systems to provide an array of varied adaptive expressions. Many of Goldschmidt's thoughts which so horrified evolutionists of his day were quite bland and unexciting but were embedded in startling phrases.

The use of existing integrative processes in evolutionary changes is sometimes dramatically seen in neotenic evolution. Common examples include salamander populations whose members, instead of transforming to the adult stage of ontogeny, retain the juvenile adaptive form though they attain adult size and sexual maturity. The juvenile adaptive plan is obviously biologically "coherent" and offers an adaptive alternative in evolution [6].

It must be true that some types of systems, or particular components within systems, are more adaptable than others. These systems or components have a greater potentiality in adjusting to changing environments. Some of these could compensate, conceivably, for adaptive imperfections in other parts of the individual. The behavioral machinery is often likely to be one such compensating system which, especially in higher animals, can make the best use of the morphological equipment it governs. There is immediate evidence that this is so. Observations on injured animals confirm their ability to learn to do the best with what they have, and some of them seem to do almost as well as uninjured individuals [7]. Among humans there are numerous examples of persons who conquered what seem to have been overwhelming handicaps to excel as great athletes and/or great minds, and these examples extend from such a personality as that of Isaac Newton to one as different as that of Abraham Stoker [8].

The compensating value of any given adaptable system, like the nervous system, is neither limitless nor a factor in all evolutionary changes involving the system. Maynard Smith (1951) notes that the structure of the flight mechanism in ancient birds, pterosaurs, and early insects was such as to ensure high stability in the air. In this way these animals resembled commercial transport planes, which are very stable, but were unlike fighter planes, whose design sacrifices stability for great maneuverability. Only later did instability evolve, as the nervous system and sense organs became more sophisticated. It seems that the aerodynamic features for stable flight in these examples evolved rapidly, faster than the nervous control, and in a sense the structure of wing, musculature, body, and tail compensated for the less adaptable nervous system. However, bats evolved instability very rapidly, and this might have been due to the great adaptive flexibility of the mammalian brain. The adaptive value of a system that can compensate for others must often be huge, for not only can such systems provide adjustments to changes within themselves but can give to other systems some of the same advantages.

The compensating phenomenon that can exist in anatomical structures has its counterpart in the coadaptation of genes in a population (see Wallace, 1968). An individual's dependence upon the activities and interactions of its constellation of genes requires the genes to be compatible. Compatibility can imply a bit of compensation, shown especially by Wallace's radiation experiments (e.g., Wallace, 1956) on *Drosophila*. These revealed that populations can contain many potentially deleterious genes but usually, when in combination with themselves and other genes, there are compensating effects that nullify the harmful properties. Because of the selection on individuals, the population's pool of genes is built up as a coadapted complex, where the frequencies of the genes are such that the chances of combining, through mating, the "wrong" genes in a single individual are tiny.

Integrated relationships among components in a system can be demonstrated through experiment, but also through measurement and statistical comparison. Parts of animals (especially vertebrate bones or teeth, molluscan shells, or other hard, easily measured structures) are often used [9]. The comparison is aimed toward determining how much one part correlates with another. For instance in a collection of skeletons from a population of, say, grass frogs, a number of bones are measured and correlations between all possible pairs of kinds of elements are determined. It might be learned that the lengths of several hindlimb bones are closely correlated, which simply means that the growth of shank, thigh, and so forth are somehow interrelated rather precisely; if you knew the shank length, you could closely estimate the thigh length with considerable confidence.

There is a reasonable expectation that components which correlate highly are those which must functionally cooperate in some kind of adaptive system. Cooperating components need a precise, operational "fitting together." High correlations are interpreted as evidence that the involved components mutually adjust. Adjustment capacity is adaptively sensible if parts are obligated to rather exact, functional cooperation. The fact of the correlations, however, does not protect the adjustment interpretation from criticism. One could critically argue that in wild populations those individuals whose interworking components are unmatched are quickly weeded out by selection, leaving behind the others to be collected and studied by investigators who can then happily discover the correlations. But studies on animals raised in captivity, and on rich collections of fossil bones where, presumably, there is less bias by the investigator in choosing individual animals with greater survivorship capabilities, support the adjustment hypothesis. Thus it appears that correlated features are held in their relationship to one another by a developmental program which guides their growth and form through mutual interaction.

Not very long ago Gould (1967) performed a correlation analysis on the bones of an extinct reptile group, the pelycosaurs. The pelycosaurs contained a number of species and genera, one of which is the rather familiar "sail-backed" form, *Dimetrodon,* pictured in Figure 48 (by no means were all pelycosaurs endowed with a "sail"). Gould's data on bone measurements came from a colossal monograph on the pelycosaurs, written some years earlier by Romer and Price (1940). These early investigators concluded that the adaptive key to evolutionary radiation of skull form in the pelycosaurs was an increase in the length of the face, with concomitant enlargement and specialization of the teeth. Gould was able to show that the correlations among characters relating to facial length and to dentition were strong enough to easily support the view of Romer and Price (see Figure 49). Of at least equal interest, though, were the results of Gould's examination of limb-bone proportions.

Particularly involved in this part of the analysis were length/width ratios of the limb bones. The ratios point directly to a biomechanical and dimensional problem

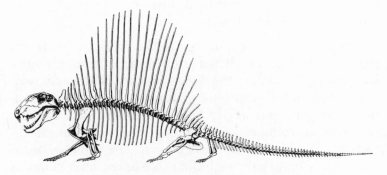

Skeleton of the Permian, sail-backed, pelycosaurian reptile, **48**
Dimetrodon. (From Romer and Price, 1940.)

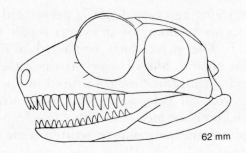

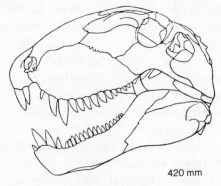

49   Two pelycosaurian skulls compared to show differences in facial length and the correlated differences in tooth characteristics. (From Gould, 1967.)

recognized at least as long ago as the early 1600s. It was then that Galileo noted that in a series of roughly similarly shaped animals, the larger ones would need relatively thicker leg bones compared to the smaller examples. The mass and weight of an animal are each proportional to its volume, itself proportional to its linear dimensions raised to the third power. The cross-sectional area of any body part, including that of the limb bones, is proportional only to the second power of the linear size. Since it is upon this cross section that the weight of the animal rests, larger animals tend to overweigh the capacities of their legs. Or I should say they would were it not for the fact that their limb bones are thicker for their length than in smaller animals.

The proportionately thicker bones of larger pelycosaurs can be appreciated by Gould's illustration (see Figure 50), where he compares shapes of thigh bones from three different-sized pelycosaurs by drawing them as if they were all three the same length. Length and width measurements for the bones can be plotted on logarithmic coordinates to depict the allometric relationships [10] involved. The three graphs in Figure 51 are from Gould's paper and show that the femur length and body length are related in a nearly one-to-one ratio (α = 0.96). Femur widths, contrastingly, are significantly great-

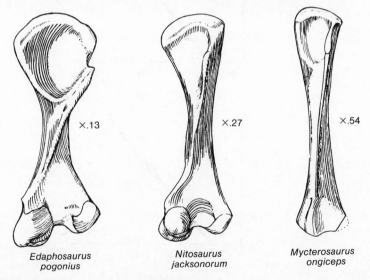

×.13

×.27

×.54

Edaphosaurus pogonius

Nitosaurus jacksonorum

Mycterosaurus ongiceps

Femur bones of three different-sized pelycosaur species. The **50** bones are drawn to the same length, and the reduction factor for each is indicated. (From Gould, 1967.)

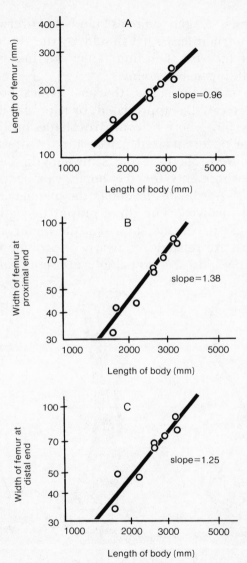

51 Double logarithmic plots of pelycosaur femur dimensions versus body length. (From Gould, 1967.)

er for bigger animals ($\alpha \gg 1.0$). One could expect that this important relationship of width to length would show itself as a strong correlation between the two dimensions. But it does not. The width correlates instead with the estimated body weights of the animals. This is far more biologically sensible than is a strong correlation between length and width. From the standpoint of controlling the growth in width, the length would be relevant only as an imperfect indicator of body weight. It is better to control bone width directly by weight than indirectly by bone length, because it is the weight to which the width must answer.

Adaptive systems, as shown in these examples where component parts can respond to relevant factors, perhaps never would have evolved were it not that individuals face a variety of experiences in their lives [11]. Some of these experiences are, in a sense, "predictable"; it can be known that if an individual tree snail lives long enough, its shell will become heavier. But it cannot be known whether the tree occupied by the snail will be struck by lightning.

Whether the variation in the life experiences of an individual are predictable or not, the fact of the variation is the source of selective pressures favoring integrating, responsive patterns. Yet the kind of pattern and its evolutionary implications can depend somewhat upon the relative predictability, or lack of it, of the conditions relating to it.

There certainly is some predictability in developmental sequences of animals. In pelycosaurs or other vertebrates, clearly something has gone wrong if an individual growing to greater lengths does not become heavier. The cross sections of supporting bones might then be controlled by bone length, which, in turn, roughly estimates the expected weight. Perhaps some kinds of developmental systems work this way. But better ones will adaptively acknowledge that things often do "go wrong," at least a little bit, and a more relevant adaptive response will associate bone cross-sectional area directly with weight, as we have seen in Gould's pelycosaurs.

In the evolution of these systems there must be a beginning point where a change in one component can, thanks to a novel genetic change, exert an appropriate control over some other component. Let us say that we are able to trace the evolution of such a system in terms of two components, X and Y. At the start, these two parts coordinate functionally to produce a workably adaptive effect except when, for any imaginable reason, the precise form of component X is altered. Suppose that a slightly altered form of X is X', a greater alteration in the same direction is X'', and so forth to X''', X'''', etc.

If Y's shape or size is independent of X, an altered X can be combined with an unaltered Y to give such pairings as X' with Y, or X'' with Y. These will be considered inadaptive. The evolution of an appropriate pattern of interacting responses must be relatively difficult, but it clearly has happened. Any heritable change that promotes an alteration of Y to Y' when X is expressed as X' can be highly favored by selection whenever the combination X' with Y' is as selectively fit, or nearly so, as X with Y.

The type of deviation that X might undergo will frequently not be restricted to a single alternative expression, such as X' alone, but a continuous (although of course limited) range of possible expressions, such as X', X'', and X'''. This will clearly be appreciated when simple growth enlargements are considered, but the expectation of a continuum could be justly held in many other circumstances, too. A well-integrated system ought to produce functionally matching combinations such as X with Y, X' with Y', X'' with Y'', and X''' with Y'''.

It is asking quite a lot of a new genetic change to produce a phenotypic response pattern so that one component adjusts properly to the vagaries of another. The relative improbability of the event, as compared with less spectacular evolutionary changes, must expose it to the restrictions that have been discussed beginning with Chapter Four. This is not to say that the event *had* to occur within that restrictive framework, but only that the likelihood that it did so looms larger with more unlikely events.

There is no necessity to imagine that, in the earliest stages in the evolution of an interacting system, there must be a great precision in the matching of components. A requirement of great precision makes the earliest steps hard to imagine at all. Components X and Y in their unaltered states might be precise, but the pair X' and Y', when changes in X invoke changes in Y, might possess only a very rough adaptive harmony at first. The selective advantage could still be very high, especially if the crude $X'Y'$ pair is much better than $X'Y$, and if in the evolving population the X component is frequently altered. It is possible, in fact, that any responsive adaptive system that *can* be formed by several components will possess a special potentiality which, because of its flexibility, might quickly become a dominant theme in the animal's adaptive repertoire.

When selection favors a response pattern, the interacting associations will frequently reflect, perhaps very crudely at first, a transcription of some physicomathematical law into biological terms. This strikes me as a very major point. In the limb-bone example, if X is the weight of the animal and Y the bone width, the adjustment of width to weight can be stated, in the symbols of our very simple model, for the range of increasing weight such as X with Y, X' with Y', and so on. But what we are in truth dealing with here is an expression of a general physical law, expressed in the "language" of biological growth parameters rather than in the written language of a physics text. This embodiment of physical equations into the formation of an individual's developmental program of response patterns allows the remarkable preadaptive abilities which have lent such speed to the evolutionary rates of many groups.

In a population of moderately sized animals, the need to possess a system of appropriate responses between weight and limb width can result in a biological organization that "recognizes" and "appreciates" pertinent physical laws. It is a relatively simple matter to evolve, from a species of moderate animals, a larger one as long as the factors of recognition and appreciation are intact. Not only developmentally related systems, but also those not strongly tied to progressive ontogenetic programs, can often show such capabilities. Any evolving system recognizing physical realities can often be extrapolated to suit functional demands of the future under new, and so far untested, selection pressures. There is thus, in many integrated systems, the preadaptive *power of extrapolation* [12]. In a real sense the evolution of any complex system in response to current circumstances is a great bid for evolutionary success in yet unborn species in yet unseen environments. As long as the physical rules of the universe do not change, the future environments are only unseen, not unknown.

An extrapolatable system does not have to embody physical laws completely and perfectly to be important in evolution. When imperfections exist, they might set ultimate limits for future evolution by extrapolation [13], although other factors could impose restrictions as well [14].

The repeated emphasis of all these arguments is that a relatively meager number of fundamental patterns underlie a great animal diversity. The basic patterns were relatively difficult to evolve, and most of them might have come early in the history of life, as some evidence suggests. Within broad groupings of animals the apparent array of greatly differing forms might in a sense be misleading. Differences are there, but their evolutionary distinctness in pattern and program is possibly less than our esthetic perceptions easily allow. The greatest feat of the evolutionary process might lie in its having completed its great feats. In this accomplishment it has done almost all that could be done for the creatures of the past, and, inadvertently, for those to come later in whatever world the future designed.

I recognize that this view will not spark joyful celebration in those who need to view their own species as some momentous, heavenly innovation among the poor, dumb creatures of the world. Human gear, in my view, is an extrapolation of the equipment evolved for some long past ancestor. I see nothing in mankind to suggest otherwise. Nor, I must confess, do I see why so many authors sound such exalting tones when discussing human evolution. Possibly they are ignorant of the blood-spattered, tear-stained pages of human history. Or possibly from that history they have singled out a few wise and loving characters who have existed, and who exist, and then attempted to hide the great welter of cruelty and intolerance behind these great principals. I urgently hope the exalters are basically right about human nature, but we most assuredly lack evidence so far.

It is not that I am blind to what potentially seems present in all humans—an esthetic sense, a curiosity, at least a hint of some selflessness. It is that these seem to me, in general, not developed to the extent, nor in the properly balanced combinations, to raise populations of humans to the moral, honest, considerate, and intellectual heights which these same humans define in complex ways as "good." If this "good" is ever to be achieved, it will occur only by man understanding himself in terms far more honest and realistic than those of his present self-perceptions. All evolutionary studies are about man, for all of them touch the reality of what he is, and what he might be.

NOTES

1. Except in some instances involving genetic drift (see Chapters 2 and 4).

2. It is likely that few disturbed systems, even when repaired by canalization, are exactly as adaptively good as others which have not been disturbed. The selective value of canalization is simply that a repaired system is better than one which can effect *no* repair of its damage.

3. See Chapter 7.

4. I have discussed some difficulties in evolving improbable adaptive systems in Chapter 4.

5. See also Dubrul (1950) and Sawin, Fox, and Latimer (1970).

6. Neotenic evolution is dealt with extensively by De Beer (1958).

7. See Chapter 5.

8. "Bram" Stoker, the author of *Dracula*.

9. For example, Schaeffer (1956) and Olson and Miller (1958) [but see Bock (1960)], Van Valen (1965), Kurtén (1967), and Gould (1967)].

10. See Chapter 7.

11. Waddington (1968) provides the beginnings of a good mathematical model showing that the selective advantages of canalization depend upon a nonconstant environment. The approach of the model suggests some compatibility with the population model of Levins (e.g., 1968), where he treats the effects of different selection factors within a habitat.

12. It is possible to provide a terribly crude illustration of this notion. Suppose that we are regarding an animal throughout its growth. Let its mass at any time be M and its standing weight be proportional to it. We are going to worry about the diameter of its leg bones. Of course the leg bones must do more than bear its "standing weight," since the animals will be prancing, jumping, and, on occasions, accidentally falling. The effective load its leg bones ought to bear is a "reasonable maximum" force that it can expect to bear during any of these activities. This effective load, or effective weight, would not unreasonably be related to the *momentum*, MV (mass × velocity), an animal would have when it hit the ground after a good jump or bad fall. The effective weight could also be related to the strength of its muscles. For instance, in a moment of confusion during a bad fall, an ill-timed strenuous contraction could further stress the limb bones. Muscular strength is roughly proportional to the cross-sectional area of the involved muscles. Thus the effective weight, X, could be estimated as

$$X = MVC_1 + M^{2/3}C_2.$$

The terms C_1 and C_2 are appropriate constants. And since the mass of the animal is related to the cube of its linear dimensions, and the cross-sectional area of pertinent muscles is related to the square, the relationship between muscle cross section and mass can be approximated by $M^{2/3}$. If we assume that animals of a certain type (e.g., all pelycosaurs) can tumble from the same maximum heights regardless of size, V also becomes a constant in the equation.

The leg bones must have a cross-sectional area sufficient to distribute the force of effective weight without damage. If each leg bone must bear the force equally, its diameter, Y, must satisfy

$$\frac{\pi(\frac{1}{2}Y)^2}{X} = K, \quad K = \text{constant}.$$

The solution for Y is

$$Y = \sqrt{\frac{4KX}{\pi}},$$

$$Y \propto \sqrt{MVC_1 + M^{2/3}C_2}, \quad \text{constants: } V, C_1, C_2;$$

$$\propto \sqrt{X}.$$

If this relationship between X and Y accurately represented the necessary conditions for a given group of animals, it would exist on the page as a physicomathematical expression. But if the integrative processes of the animals responded according to the formulation, the integrations themselves would be an equation, not in conventional mathematical terms to be written down on paper, but in the terms of biological parameters of growth and form. In an idealized case, the biological equation would show appropriate changes in Y for any given change in X. But the biological equation, still thinking of this in the idealistic realm, would state the necessary conditions, within the ontogenetic range, for animals of any size—or mass—and would, in effect, state a general, physical reality whose application to many situations is enormous unless the laws of physics change significantly.

The physicomathematical rule could be expressed in the allometric growth form as

$$Y \propto X^{1/2}$$

or

$$Y = bX^{1/2}$$

with b designed as an appropriate constant. Within the baby pelycosaur there must exist some program of biological organization that will serve it as it grows to adulthood, and which does so through a transcription of the equations of physics. If these "equations" are sufficiently general, they will anticipate future conditions, permitting and even promoting future evolutionary changes. Thus future new forms will often be extensions from ancestral types whose basic adaptive responses are extrapolatable, hence in a real sense anticipatory.

Clearly this example is merely a verbal diagram of the notion of the *power of extrapolation*. The factors that enter our consideration of "effective weight" are only roughly surmised for illustration, and the exactitude of the mathematical relationships, as they exist within a creature's developmental program, is probably overdone in this example. Nevertheless, the evidence that extrapolatable or anticipatory systems play a great role in evolutionary radiations is becoming harder to ignore.

13. As an example, consider the dimensional relationships of two parts X and Y. Let us say that they relate mechanically to one another such that when X is one unit in size, Y must be one also. But as we continue to imagine X increasing to two, to three, to four, and so forth, units in size, we observe Y's own required increase to display the relationship with X:

If X is this size	1	2	3	4	5	6	$\cdots$	$X(n)$
Y must be	1	2	4	8	16	32	$\cdots$	$2Y_{X(n)-1}$

The relationship simply states that each time X increases by one unit, Y doubles in length from its immediately previous value [that Y value corresponding to $X (n) - 1$, where n indexes the present size of X]. Now if this simple relationship defines the *required* functional relationship between X and Y, we should look at the *actual* circumstances regarding X and Y in an animal.

The animal I have in mind is one whose X size begins at one unit and increases to three units before the animal loses reproductive ability. We note that Y maintains its proper relationship so that as X increases Y doubles each time, to give

X	1	2	3
Y	1	2	4

In subsequent generations selection has favored a larger X, which then reaches four units in size, and Y will double to a size of eight units. However, in this example, that is as far as things will go. Selection cannot favor a further increase of X to five units because at $X = 5$, $Y = 4$. The relationship is *not* based on the equation $Y(n) = 2Y(n-1) = 2Y_{X(n)-1}$. It is based instead on $Y(n) = [Y(n - 1) + X(n)] - [X(n) - Y(n - 1)]^2$. This equation gives the pattern

X	1	2	3	4	5	6	7
Y	1	2	4	8	4	6	12

If in the evolution of the relationship between X and Y the size of X never exceeded four units, the "wrong" quantitative association would have been as selectively favored as the "right" one. The only difference is one that selection was powerless to do anything about it, and that is the inflexibility of the relationship for future evolution.

14. The variety of systems in an animal could, at some extreme levels of extrapolative evolution of one or more of the systems, produce interference. Overcoming the interference might be impossible without evolutionary changes that integrate the *systems* in a way resembling the integra-

tion of components within each system. Moreover, some physical relationships, even if accurately operative in animals, have limits, such as that between mass and surface, which must impose serious restrictions at the extremes of growth.

References

Alexander, R. D., and W. L. Brown, Jr. 1963. Mating behavior and the origin of insect wings. *Occas. Papers, Mus. Zool., Michigan 628:* 1–19.

Alexander, R. M. 1968 *Animal mechanics.* Sidgwick & Jackson, London.

Alexander, R. M. 1971. *Size and shape.* Edward Arnold, London.

Anderson, J. D., D. D. Hassinger, and G. H. Dalrymple. 1971. Natural mortality of eggs and larvae of *Ambystoma tigrinum. Ecol. 52:* 1107–1112.

Andrewartha, H. G., and L. C. Birch. 1954. *The distribution and abundance of animals.* University of Chicago Press, Chicago.

Banks, N. L. 1970. Trace fossils from the late Precambrian and Lower Cambrian of Finnmark, Norway. In *Trace fossils* (T. P. Crimes and J. C. Harper, eds.), pp. 19–34. J. Geol. Spec. Issue 3.

Bassett, C. A. L. 1971. Biophysical principles affecting bone structures. In *The biochemistry and physiology of bone,* Vol. 3 (G. H. Bourne, ed.), pp. 1–76. Academic Press, New York.

Berlese, A. 1913. Intorno alle metamorfosi degle insetti. *Redia 9:* 121–138.

Bock, W. J. 1959. Preadaptation and multiple evolutionary pathways. *Evol. 13:* 194–211.

Bock, W. J. 1960. A critique of "morphological integration." *Evol. 14:* 130–132.

Bock, W. J. 1970. Microevolutionary sequences as a fundamental concept in macroevolutionary models. *Evol. 24:* 704–722.

Boyd, T. G., W. A. Castelli, and D. F. Huelke. 1967. Removal of the temporalis muscle from its origin: effects on the size and shape of the coronoid process. *J. Dental Res. 46:* 997–1001.

Brink, A. S. 1956. Speculations on some advanced mammalian characteristics in the higher mammal-like reptiles. *Paleontol. Africana 4:* 77–96.

Britten, R. J., and E. H. Davidson. 1971. Repetitive and non-repetitive DNA sequences and a speculation on the origins of evolutionary novelty. *Quart. Rev. Biol. 46:* 111–138.

Brough, J. 1958. Time and evolution. In *Studies on fossil vertebrates* (T. S. Westoll, ed.), pp. 16–38. University of London, London.

Brown, W. L., and E. O. Wilson. 1956. Character displacement. *System. Zool. 5:* 49–64.

Brues, A. M. 1964. The cost of evolution versus the cost of not evolving. *Evol. 18:* 379–383.

Calder, R. 1968. *The evolution of the machine* (American Heritage, for the Smithsonian Institution). Van Nostrand, Reinhold, New York.

Camin, J. H., and P. R. Ehrlich. 1958. Natural selection in water snakes (*Natrix sipedon* L.) on islands in Lake Erie. *Evol. 12:* 504–511.

Cloud, P. E., Jr. 1968. Pre-metazoan evolution and the origins of the Metazoa. In *Evolution and Environment* (E. T. Drake, ed.), pp. 1–72. Yale University Press, New Haven.

Cohn, M. 1970. Anticipatory mechanisms of individuals. In *Control processes in multicellular organisms* (G. E. W. Wolstenholme and J. Knight, eds.), pp. 255–303 (Ciba Foundation Symposium), J. & A. Churchill, London.

Cole, L. C. 1954. The population consequences of life history phenomena. *Quart. Rev. Biol. 29:* 103–137.

Collette, B. B. 1961. Correlations between ecology and morphology in anoline lizards from Havana, Cuba and Southern Florida. *Bull. Mus. Comp. Zool. 125:* 135–162.

Crow, J. F. 1970. Genetic loads and the cost of natural selection. In *Mathematical topics in population genetics.* K. Kojima (ed.), pp. 128–177.

Crow, J. F., and M. Kimura. 1970. *An introduction to population genetics theory.* Harper & Row, New York.

Currey, J. D. 1968. The adaptation of bones to stress. *J. Theoret. Biol. 20:* 91–106.

Davis, D. D. 1949. Comparative anatomy and the evolution of vertebrates. In *Genetics, paleontology and evolution.* (G. L. Jepsen, E. Mayr, and G. G. Simpson, eds.), pp. 64–89. Princeton University Press, Princeton, N. J.

De Beer, G. R. 1958. *Embryos and ancestors,* 3rd ed. Clarendon Press, Oxford.

Deevey, E. S. 1947. Life tables for natural populations of animals. *Quart. Rev. Biol. 22:* 283–314.

DeJong, H. J. 1968. Functional morphology of the jaw apparatus of larval and metamorphosing *Rana temporaria* L. *Netherlands J. Zool. 18:* 1–103.

Derry, T. K., and T. I. Williams. 1960. *A short history of technology.* Oxford University Press, New York.

Devillers, C. 1965. The role of morphogenesis in the origin of higher levels of organization. *System. Zool. 14:* 259–271.

DuBrul, E. L. 1950. Posture, locomotion and the skull in lagomorpha. *Amer. J. Anat.* 87: 277–313.

DuBrul, E. L., and D. M. Laskin. 1961. Preadaptive potentialities of the mammalian skull: an experiment in growth and form. *Amer. J. Anat.* 109: 117–132.

Dunn, E. R. 1942. Survival value of varietal characters in snakes. *Amer. Natur.* 76: 104–109.

Eden, M. 1967. Inadequacies of neo-Darwinian evolution as a scientific theory. In *Mathematical challenges to the neo-Darwinian interpretation of evolution* (P. S. Moorhead and M. M. Kaplan, eds.). Wistar Inst. Sympos. Monograph 5, pp. 5–19, 109–112.

Emlen, J. M. 1970. Age specificity and ecological theory. *Ecol.* 51: 588–601.

Emlen, J. M. 1973. *Ecology: an evolutionary approach.* Addison-Wesley, Reading, Mass.

Enlow, D. H. 1963. *Principles of bone remodeling.* Charles C Thomas, Springfield, Ill.

Eshel, I., and M. W. Feldman. 1970. On the evolutionary effect of recombination. *Theoret. Pop. Biol.* 1: 88–100.

Ewens, W. J. 1970. Remarks on the substitutional load. *Theoret. Pop. Biol.* 1: 129–139.

Falconer, D. S. 1960. *Introduction to quantitative genetics.* Ronald Press, New York.

Felsenstein, J. 1971. On the biological significance of the cost of gene substitution. *Amer. Natur.* 105: 1–11.

Ferguson, E. S. 1967. The steam engine before 1830. In *Technology in western civilization,* vol. 1: pp. 245–263. (M. Kranzberg and C. W. Pursell, Jr., eds.) Oxford University Press, New York.

Fisher, R. A. 1958. *The genetical theory of natural selection,* 2nd ed. Dover Publications, New York. (Original publication 1930 by The Clarendon Press, Oxford.)

Fox, R. M., and J. W. Fox. 1964. *Comparative entomology.* Van Nostrand, Reinhold, New York.

Franklin, I., and R. C. Lewontin. 1970. Is the gene the unit of selection? *Genet.* 65: 707–734.

Frazzetta, T. H. 1966. Studies on the morphology and function of the skull in the Boidae (Serpentes). Part II. Morphology and function of the jaw apparatus in *Python sebae* and *Python molurus.* *J. Morph.* 118: 217–296.

Frazzetta, T. H. 1970. From hopeful monsters to bolyerine snakes? *Amer. Natur.* 104: 55–72.

Gadgil, M., and W. H. Bossert. 1970. Life history consequences of natural selection. *Amer. Natur.* 104: 1–24.

Goin, O. B., and C. J. Goin. 1968. DNA and the evolution of the vertebrates. *Amer. Midl. Natur. 80:* 289–298.

Goldschmidt, R. 1938. *Physiological genetics.* McGraw-Hill, London.

Goldschmidt, R. 1940. *The material basis of evolution.* Yale University Press, New Haven.

Goldschmidt, R. 1946. "An empirical evolutionary generalization" viewed from the standpoint of phenogenetics. *Amer. Natur. 80:* 305–317.

Goldschmidt, R. 1955. *Theoretical genetics.* University of California Press, Berkeley, Calif.

Goss, R. J. 1965. The functional demand theory of growth regulation. In *Regeneration in animals and related problems* (V. Kiortsis and H. A. L. Trampusch, eds.), pp. 444–451. North-Holland, Amsterdam.

Gould, S. J. 1966. Allometry and size in ontogeny and phylogeny. *Biol. Rev. 41:* 347–478.

Gould, S. J. 1967. Evolutionary patterns in pelycosaurian reptiles: A factor-analytic study. *Evol. 21:* 385–401.

Gould, S. J. 1970. Evolutionary paleontology and the science of form. *Earth-Sci. Rev. 6:* 77–119.

Gould, S. J. 1971. Geometric similarity in allometric growth: a contribution to the problem of scaling in the evolution of size. *Amer. Natur. 105:* 113–136.

Grant, P. R. 1972. Convergent and divergent character displacement. *Biol. J. Linn. Soc. 4:* 39–68.

Gunter, G., and J. W. Ward. 1961. Some fishes that survive extreme injuries, and some aspects of tenacity of life. *Copeia 4:* 456–462.

Guthrie, R. D. 1969. Senescence as an adaptive trait. *Perspectives in Biol. & Med. 12:* 313–324.

Haines, R. W., 1969. Epiphyses and sesamoids. In *Biology of the Reptilia,* vol. 1, (C. Gans, A. d'A. Bellairs, and T. S. Parsons, eds.), pp. 81–115. Academic Press, New York.

Haldane, J. B. S. 1957. The cost of natural selection. *J. Genet. 55:* 511–524.

Ham, A. W., and W. R. Harris. 1971. Repair and transplantation of bone. In *The biochemistry and physiology of bone,* 2nd ed., vol. 3, (G. H. Bourne, ed.), pp. 338–400. Academic Press, New York.

Hamilton, W. D. 1966. The moulding of senescence by natural selection. *J. Theoret. Biol. 12:* 12–45.

Hampé, A. 1959. Contribution à l'étude du développement et de la régulation des déficiences et des excédents dans la patte de l'embryon de poulet. *Arch. Anat. Microscop. Morph. Exp. 48:* 347–378.

Hecht, M. K. 1965. The role of natural selection and evolutionary rates in the origin of higher levels of organization. *System. Zool. 14:* 301–317.

Hersh, A. H. 1934. Evolutionary relative growth in the titanotheres. *Amer. Natur. 58:* 537–561.

Hinton, H. E. 1948. On the origin and function of the pupal stage. *Proc. Zool. Soc. London, 116:* 282–328.

Hopson, J. A., and A. W. Crompton. 1969. Origin of mammals. *Evol. Biol. 3:* 15–72.

Hutchinson, G. E. 1962. *The enchanted voyage.* Yale University Press, New Haven.

Hutchinson, G. E., and R. H. MacArthur. 1959. A theoretical ecological model of size distribution among species of animals. *Amer. Natur. 93:* 117–125.

Huxley, J. 1932. *Problems of relative growth.* Methuen, London.

Istock, C. A. 1967. The evolution of complex life cycle phenomena: an ecological perspective. *Evol. 21:* 592–605.

Jarvik, E. 1959. Dermal fin-rays and Holmgren's principle of delamination. *Kungl. Sven. Vetensk. Akad. Handl. 6:* 1–51.

Kermack, K. A. 1967. The interrelationships of early mammals. *J. Linn. Soc. (Zool.) 311:* 241–249.

Kimura, M., and J. F. Crow. 1969. Natural selection and gene substitution. *Genet. Res. 13:* 127–141.

Kimura, M., and T. Maruyama. 1969. The substitutional load in a finite population. *Hered. 24:* 101–114.

Kimura, M., and T. Ohta. 1971. *Theoretical aspects of population genetics* (Monogr. Pop. Biol. 4). Princeton University Press, Princeton, N.J.

King, J. L. 1966. The gene interaction component of the genetic load. *Genet. 53:* 403–413.

Kurtén, B. 1957. A case of Darwinian selection in bears. *Evol. 11:* 412–416.

Kurtén, B. 1963. Return of a lost structure in the evolution of the felid dentition. *Comment. Biol. (Soc. Sci. Fennica) 26:* 1–12.

Kurtén, B. 1967. Some quantitative approaches to dental microevolution. *J. Dental Res. 46:* 817–828.

Lerner, I. M. 1954. *Genetic homeostasis.* Oliver & Boyd, London.

Levins, R. 1968. *Evolution in changing environments.* Princeton University Press, Princeton, N.J.

Li, C. C. 1955. *Population genetics.* University of Chicago Press, Chicago.

Lloyd, M., and H. S. Dybas. 1966a. The periodical cicada problem. I. Population ecology. *Evol. 20:*133–149.

Lloyd, M., and H. S. Dybas. 1966b. The periodical cicada problem. II. Evolution. *Evol. 20:*466–505.

Lotka, A. J. 1956. *Elements of mathematical biology,* 2nd ed. Dover Publications, New York.

Ludwig, W. 1950. Zur Theorie der Konkurrenz. Die Annidation (Einnischung) als fünfter Evolutionsfaktor. *Neve Ergeb. Probleme Zool. Klatt-Festschrift:* 516–537.

Maynard Smith, J. 1951. The importance of the nervous system in the evolution of animal flight. *Evol. 6:* 127–129.

Maynard Smith, J. 1968a. "Haldane's dilemma" and the rate of evolution. *Nature 219:* 1114–1116.

Maynard Smith, J. 1968b. *Mathematical ideas in biology.* Cambridge University Press, New York.

Mayr, E. 1942. *Systematics and the origin of species.* Columbia University Press, New York.

Mayr, E. 1960. The emergence of evolutionary novelties. In *Evolution after Darwin,* vol. 1. (Sol Tax, ed.), pp. 349–380. University of Chicago Press, Chicago.

Mayr, E. 1963. *Animal species and evolution.* Harvard University Press, Cambridge, Mass.

McLachlan, A. D. 1972. Repeating sequences and gene duplication in proteins. *Appendix*—Substitution frequencies in proteins. *J. Molec. Biol. 64:* 417–433, 434–438.

Medawar, P. B. 1957. *The uniqueness of the individual.* Methuen, London.

Mettler, L. E., and T. G. Gregg. 1969. *Population genetics and evolution.* Prentice-Hall, Englewood Cliffs, N.J.

Milham, W. I. 1941. *Time and timekeepers.* Macmillan, New York.

Milkman, R. D. 1967. Heterosis as a major cause of heterozygosity in nature. *Genet. 55:* 493–495.

Moorhead, P. S., and M. M. Kaplan (eds.). 1967. *Mathematical challenges to the neo-Darwinian interpretation of evolution.* Wistar Inst. Sympos. Monograph 5.

Moran, P. A. P. 1970. "Haldane's dilemma" and the rate of evolution. *Ann. Human Genet. 33:* 245–249.

Mukai, T. 1969. The genetic structure of natural populations of *Drosophila melanogaster.* VII. Synergistic interaction of spontaneous mutant polygenes controlling viability. *Genet. 61:* 149–161.

Muller, H. J. 1950. Evidence of the precision of genetic adaptation. *Harvey Lectures:* 165–229.

Murray, P. D. F. 1936. *Bones: a study of the development and structure of the vertebrate skeleton.* Cambridge University Press, New York.

Murray, P. D. F., and M. Smiles. 1965. Factors in the evocation of adventitious (secondary) cartilage in the chick embryo. *Austral. J.*

Zool. 13: 351–381.

Nei, M. 1971. Fertility excess necessary for gene substitution in regulated populations. *Genet. 68:* 169–184.

Noble, G. K., and M. E. Jaeckle. 1928. The digital pads of the tree frogs: a study of the phylogenesis of an adaptive structure. *J. Morph. Physiol. 45:* 259–292.

O'Donald, P. 1969. "Haldane's dilemma" and the rate of natural selection. *Nature 221:* 815–816.

Ohno, S. 1970. *Evolution by gene duplication.* Springer-Verlag, New York.

Olson, E. C. 1959. The evolution of mammalian characters. *Evol. 13:* 344–353.

Olson, E. C., and R. L. Miller, 1958. *Morphological integration.* University of Chicago Press, Chicago.

Osborn, H. F. 1929. *The titanotheres of ancient Wyoming, Dakota and Nebraska.* U.S. Geol. Surv. Monograph 55 (2 vols.).

Ottow, R. 1950. Die erbbedingte Osteogenesis Dyplastico-exostotica der ausgerotteten flugunfähigen Reisentauben *Pezophaps solitaria,* der Mascareneninsel Rodriguez. *K. Svensk. Vetensk. Akad. Handl., Ser. 4, 1:* 1–37.

Oxnard, C. E. 1969. Mathematics, shape and function. *Amer. Scientist 57:* 75–96.

Papanek, V. 1972. *Design for the real world.* Random House, New York.

Portmann, A. 1959. *Animal camouflage.* University of Michigan Press, Ann Arbor, Mich.

Rand, A. S. 1965. On the frequency and extent of naturally occurring foot injuries in *Tropidurus torquatus* (Sauria, Iquanidae). *Papeis Avulsos, Dept. Zool., Secr. Agric., Sao Paulo 17:* 225–228.

Rendel, J. M. 1965. The effects of genetic change at different levels. In *Ideas in modern biology.* (J. A. Moore, ed.), pp. 285–295. Doubleday, New York.

Rendel, J. M. 1967. *Canalisation and gene control.* Logos Press, London.

Rendel, J. M. 1968. Genetic control of a developmental process. In *Population biology and evolution* (R. C. Lewontin, ed.), pp. 47–66. Syracuse University Press, Syracuse, N.Y.

Rensch, B. 1959. *Evolution above the species level.* Columbia University Press, New York. (Original publication 1947 by F. Enke Verlag, Stuttgart.)

Romer, A. S. 1949. Time series and trends in animal evolution. In *Genetics, paleontology and evolution* (G. L. Jepsen, E. Mayr, and G. G. Simpson, eds.), pp. 103–120. Princeton University Press, Princeton, N.J.

Romer, A. S. 1970. *The vertebrate body.* Saunders, Philadelphia.

Romer, A. S., and L. I. Price. 1940. *Review of the Pelycosauria.* Geol. Soc. Amer. Spec. Papers 28: i–x, 1–538.

Sawin, P. B., R. R. Fox, and H. B. Latimer. 1970. Morphogenetic studies of the rabbit. XLI. Gradients of correlation in the architecture of morphology. *Amer. J. Anat. 128:* 137–146.

Schaeffer, B. 1956. Evolution in the subholostean fishes. *Evol. 10:* 201–212.

Scharloo, W. 1971. Reproductive isolation by disruptive selection: did it occur? *Amer. Natur. 105:* 83–86.

Schindewolf, O. H. 1950. *Grundfragen der Paläontologie.* Schweizerbart, Stuttgart.

Schopf, J. W. 1970. Precambrian micro-organisms and evolutionary events prior to the origin of vascular plants. *Biol. Rev. 45:* 319–352.

Simpson, G. G. 1953. *The major features of evolution.* Columbia University Press, New York.

Slobodkin, L. B. 1961. *Growth and regulation of animal populations.* Holt, Rinehart and Winston, New York.

Sloss, L. L. 1950. Rates of evolution. *J. Paleont. 24:* 131–139.

Smith, H. M., and J. C. List. 1951. The occurrence of digital pads as an anomaly in the bullfrog. *Turtox News 29:* 203.

Sondhi, K. C. 1962. The evolution of a pattern. *Evol. 16:* 186–191.

Sondhi, K. C. 1963. The biological foundations of animal patterns. *Quart. Rev. Biol. 38:* 289–327.

Stanley, S. M. 1973. An explanation for Cope's rule. *Evol. 27:* 1–26.

Stebbins, G. L. 1968. Integration of development and evolutionary progress. In *Population biology and evolution* R. C. Lewontin, ed.). Syracuse University Press, Syracuse, N.Y.

Stebbins, G. L. 1969. *The basis of progressive evolution.* University of North Carolina Press, Chapel Hill, N.C.

Sved, J. A. 1968. Possible rates of gene substitution in evolution. *Amer. Natur. 102:* 283–293.

Szarski, H. 1971. The importance of deviation amplifying circuits for the understanding of the course of evolution. *Acta Biotheoret. 20:* 158–170.

Tarlo, L. B. H. 1967. The tessellated pattern of dermal armour in the Heterostraci. *J. Linn. Soc. (Zool.) 47:* 45–54.

Thoday, J. M. 1959. Effects of disruptive selection. 1. Genetic flexibility. *Hered. 13:* 187–203.

Thoday, J. M., and J. B. Gibson. 1970. The probability of isolation by disruptive selection. *Amer. Natur. 104:* 219–230.

Thoday, J. M., and J. B. Gibson. 1971. Reply to Scharloo. *Amer. Natur. 105:* 86–88.

Thompson, D'A. W. 1942. *On growth and form.* Cambridge University Press, Cambridge.

Uzzell, T., and K. W. Corbin. 1972. Evolutionary rates in cistrons specifying mammalian hemoglobin α- and β-chains: phenetic vs. patristic measurements. *Amer. Natur. 106:* 555–573.

Valentine, J. W. 1969. Patterns of taxonomic and ecological structure of the shelf benthos during Phanerozoic time. *Paleont. 12:* 684–709.

Van Valen, L. 1960. Nonadaptive aspects of evolution. *Amer. Natur. 94:* 305–308.

Van Valen, L. 1963. Haldane's dilemma, evolutionary rates, and heterosis. *Amer. Natur. 97:* 185–190.

Van Valen, L. 1965a. Selection in natural populations. III. Measurement and estimation. *Evol. 19:* 514–528.

Van Valen, L. 1965b. The study of morphological integration. *Evol. 19:* 347–349.

Van Valen, L. 1970. An analysis of developmental fields. *Devel. Biol. 23:* 456–477.

Waddington, C. H. 1953. Genetic assimilation of an acquired character. *Evol. 7:* 118–126.

Waddington, C. H. 1957. *The strategy of the genes.* George Allen & Unwin, London.

Waddington, C. H. 1962. *New patterns in genetics and development.* Columbia University Press, New York.

Waddington, C. H. 1968. The paradigm for the evolutionary process. In *Population biology and evolution* (R. C. Lewontin, ed.), pp. 37–46. Syracuse University Press, Syracuse, N.Y.

Wallace, B. 1956. Studies on irradiated populations of *Drosophila melanogaster. J. Genet. 54:* 280–293.

Wallace, B. 1968. *Topics in population genetics.* W. W. Norton, New York.

Watson, D. M. S. 1951. *Paleontology and modern biology.* Yale University Press, New Haven.

Watt, W. B. 1972. Intragenic recombination as a source of population genetic variability. *Amer. Natur. 106:* 737–753.

Westoll, T. S. 1949. On the evolution of the Dipnoi. In *Genetics, paleontology and evolution* (G. L. Jepsen, E. Mayr, and G. G. Simpson, eds.), pp. 121–184. Princeton University Press, Princeton, N.J.

Whyte, L. L. 1964. Internal factors in evolution. *Acta Biotheoret. 18:* 33–48.

Whyte, L. L. 1965. *Internal factors in evolution.* George Braziller, New York.

Wickler, W. 1968. *Mimicry in plants and animals.* McGraw-Hill, New York.

Wiedmann, Jost. 1969. The heteromorphs and ammonoid extinction. *Biol. Rev. 44:* 563–602.

Williams, G. C. 1957. Pleiotrophy, natural selection and the evolution of senescence. *Evol. 11:* 398–411.

Williams, G. C. 1966. *Adaptation and natural selection, a critique of some current evolutionary thought.* Princeton University Press, Princeton, N.J.

Wilson, E. O., and W. H. Bossert. 1971. *A primer on population biology.* Sinauer Associates, Sunderland, Mass.

Wright, S. 1932. The roles of mutation, inbreeding, crossbreeding, and selection in evolution. *Proc. 6th Internat. Congr. Genet. 1:* 356–366.

Wright, S. 1956. Modes of selection. *Amer. Natur. 90:* 5–24.

Author Index

Subject Index

About the Book

The text of this book is set in V.I.P. Baskerville, a modification of an English typeface designed in the 18th century by John Baskerville. The display type is set in Bulmer. The book was designed by Joseph J. Vesely, set by David E. Seham Associates, printed by Bloomsburg Craftsmen, and bound by Haddon Craftsmen, Inc.